AF880073

The Physical Testing of Paper

Roman E. Popil

First Published in 2017

This book was originally published by

Smithers Group Company
Shawbury, Shrewsbury, Shropshire, SY4 4NR, United Kingdom
Telephone: +44 (0)1939 250383 Fax: +44 (0)1939 251118
http://www.polymer-books.com

A catalogue record for this book is available from the British Library.

ISBN: 978-191024-292-6 (Softback)
978-191024-293-3 (ebook)

Typeset by Argil Services

Preface

The '*Physical Testing of Paper*' is a broad term which covers the measurement of physical properties of various papers, paperboards, corrugated boards and corrugated board boxes. Perhaps 'wood-fibre based laminate composite structures' could be substituted as being more encompassing than the generic 'paper' but that would make a rather long title. Quite often, this vast subject receives only cursory treatment in institutions where papermaking is taught as an ancillary course to a chemical engineering curriculum concerning the manufacturing processes of paper. An unfortunate and misleading impression encountered here is that paper testing is simply a matter of pressing a button on some ready-made instrument and recording the resulting number and is therefore undeserving of much attention or focus – not so! I have countered this somewhat condescending oversimplification by stating in retaliation that chemistry by contrast, appears to be, to the casual hallway passer-by, simply mixing various chemicals in a beaker while wearing a white lab coat! This amusing discourse continues to this day with my colleagues with the results being as can be imagined.

This book arises from my experience gained from entering the paper industry with an experimental physics background some decades ago. In this approach, test methods that have been documented as standards are subject to scrutiny, interpretation, improvement and improvisation. There is the requirement and desire to understand what is going on from a fundamental mechanics view when a paper sample test piece undergoing a test is being pulled or torn, bent or poked. Repeatedly, I find myself required to explain in layman's terms to my testing lab clients what it is that is going on in their submitted paper samples that disallows them to be nicely printed, written on without ink blobbing onto the next page, glued together so the assembled carton does not fall apart or folded without cracking. There is meaning and satisfaction in linking the results of various different tests to diagnose some end-use issue such as gluability, printability or corrugated box storage longevity. The significance of the results obtained from various testing methods is what is emphasised in this volume rather than the details of the test methods, which are documented elsewhere.

In this book I have focused on the basic physical tests that have demonstrated interrelationships and solved problems over the years. Many more different tests are occasionally requested but have not resulted in providing significant insight and are not included here for brevity. These include ZD out-of-plane tensile, Scott Bond, folding resistance, surface abrasion resistance, surface friction, surface electrical resistivity, water absorption and water resistance in its many forms. I refer the interested reader wishing to delve further to the references cited below for more details of the testing methods, as well as the other listed texts for the paper physics associated with paper testing.

The book is written from a personal view based on my initial experience as an instrument developer for a newsprint production company in Canada, followed by similar activity for a sensor scanner manufacturer in California. Here, new measurement methods and techniques were developed to provide unique paper property characterisation that was not otherwise available by conventional testing. Since 2003, I have managed the Physical Analysis Lab at the Institute of Paper Technology and Science in Atlanta, GA, USA, now renamed the Renewable Bioproducts Institute as part of the Georgia Institute of Technology in Atlanta, GA, USA. Here, I have encountered various testing requests from a variety of paper manufacturers and end-users to address a wide variety of quality issues. Although much of the activity still involves the apparent tedium of repeatedly pressing a button on an instrument and recording the resulting numbers, there is excitement, which I convey to my clients and any students within earshot, that lies in realising what all the numbers mean and how they can be used together to solve a particular problem.

Thus, I invite the reader to share the joy and excitement of paper testing by taking the time to explore some of the fundamentals of paper physics related to property measurements, to take a curious investigative and skeptical approach to the subject and finally, realise how data obtained from testing can be used to tell a story that is not only informative, but also gratifying.

Bibliography

1. *TAPPI Testing Methods*, Tappi Press, Atlanta, GA, USA.

2. *Handbook of Physical Testing of Paper*, 2nd Edition, Eds., R.E. Mark, C.C. Habeger, Jr., J. Borch and M.B. Lyne, Marcel Decker, Inc., New York, NY, USA, 2002.

3. K.J. Niskanen in *Paper Physics Book 16 of Papermaking Science and Technology*, Fapet OY, Helsinki, Finland, 1998.

4. J.E. Levlin and L. Söderhjelm in *Pulp and Paper Testing*, Fapet Oy, Helsinki, Finland, 1999.

5. J.D. Peel in *Paper Science and Manufacture*, Angus Wild, Vancouver, Canada, 1999.

6. G.A. Smook in *Handbook for Pulp and Paper Technologists*, Angus Wild, Vancouver, Canada, 2002.

Contents

Errata

Print copies of this book issued prior to October 2019 contain some inadvertent errors to which the author has been kindly alerted by several readers, notably Abdul Wahab, Tomas Nordstrand and others. These are corrected in this current text and are listed below for reference.

1) **Equation 1.2** subscripts are corrected to read:

   $SCT_{50\%RH} = SCT_{\%M} \times 0.531e^{0.089(\%M)}$ and similarly, so is the equation embedded in Figure 1.2

2) Page 17, 3rd paragraph, line 6, now reads: … "the wire side surface often may have lower strength."

3) Page 23, 3rd paragraph, 2nd line, now reads: "machine-made paper (MD) or the cross direction of paper (CD),"

4) Page 47 **Figure 3.6**, horizontal axis units are kN/mm.

5) Page 48, **Figure 3.7** horizontal axis units are kN/mm.

6) Page 71, 2nd line from top reads "to the equation (which neglects the shear contribution to bending): "

7) Page 71, Equation 4.12 the length L is to the 3rd power and so is:

$$D = PL^3/48wy \qquad (4.12)$$

8) Page 90, 2nd paragraph, line 6 reads "The top side, when fines interfere with fiber bonding, can be somewhat weaker than the wire side."

9) Page 98, **Table 5.2,** the board caliper in mm for A- flute is 5.2 , the slenderness ratio is 33.8, for C- flute the caliper is 4.3, slenderness ratio 30.7, for E-flute, caliper is 1.2 and slenderness 144

10) Page 121, **Table 5.4**, under "Board description" heading 2nd column, line 6 reads "Lab-made C-flute", line 7 reads "Lab-made C-flute 2nd set".

1 Introduction –What is Paper?

One of my former physics professors upon hearing that I had started working in the paper industry as a research scientist, dismissed all the alleged complexities of paper structure with a wave of the hand and saying that paper is 'just reconstituted wood'. Those three words, although being a typically glib oversimplification, are actually true and are helpful to understand the nature of those things we usually call paper. However, it was indeed an 18th century physicist, Antoine de Reamur [1], who suggested in 1719 that his observation of the maceration of wood by wasps to form their nests could be replicated by humans to make something similar and useful. At the time, paper was made from disintegrated cotton rags which were increasingly in short supply as demand increased. Wood is wood and paper is paper and, other than some wallpaper products that are deliberately printed to look like wood, they do not look the same. The similarity, however, is that all paper consists of conglomerates of bonded fibres, all of which were once the living cells of a tree or plant [2]. These fibres are able to form paper because they consist of cellulose, which is hydrophilic, and so are able to absorb and retain water. The individual fibres are also typically small, widths are approximately 30 μm, about one-third of the width of a human hair, and 1 or 2 mm long, sometimes longer or shorter depending on the wood or plant species [3]. If a piece of paper is torn, these fibres can be seen by the eye sticking out along the edge of the tear. Once the fibres have been separated from the plant, by mechanical or chemical means, dispersed in water and then strained, the hydrophilicity of the cellulosic surface draws the fibres together and as the strained mat dries, enough water becomes entrained to form hydrogen bonds between the individual fibres, forming a matrix called paper [4]. If the process is reversed, i.e., a copious amount of water is added to paper, the water/paper mix is sufficiently agitated to disperse the paper into separated fibres and the resulting slurry is strained and then dried, the mat will reform [5] albeit with some irreversible changes to the fibres and loss of sheet strength. Hence, the rather dismissive moniker of 'reconstituted wood' is justified using this wet-laid strained slurry process by which most paper is made. Thinking of paper as consisting of agglomerations of fibres can be visualised by the analogy of fallen pine tree needles on a road, as shown in **Figure 1.1**.

Figure 1.1 A naturally occurring paper analogy can be mats and clumps of intertwined pine 'straw' laying on a road. Repeated traffic has aligned the 'straws' somewhat in a left to right direction in this photograph

Here, the aspect ratio of fibre width to length is similar to paper magnified to a scale of 100×. Thus, paper consists of a mat of cellulose-containing fibres that are bonded together *via* the hydrogen-bonding mechanism occurring through fibre contact. On the scale of approximately 1 mm, which can be easily observed through a low-power microscope, the surface of paper in transmitted light can be said to resemble a plate of cooked fettuccine pasta. The wood fibres, which originated as hollow tubes in the papermaking stock dispersion, have become transversely compressed and pressed dry in the paper. They have thus collapsed to form flat ribbons with a dog bone-shaped cross-section, as may be gleaned from the 2-photon fluorescence scanning laser confocal shot [6] of **Figure 1.2** and the optical cross-section of newsprint shown in **Figure 1.3**. Chemically prepared paper, referred to as kraft, (German for 'strong') forms flattened ribbon fibres, obvious in **Figure 1.3**, where the newsprint shown in cross-sections consists of flattened kraft fibres as well as thick-walled uncollapsed mechanical pulp fibres. Thick-walled fibres are also attributable to latewood, a result of fibre walls thickening as tree growth slows down during colder temperatures, creating the rings that we see when looking at the cross-section of a cut tree.

Figure 1.2 A confocal 2-photon fluorescence scanning laser microscopy shot showing a layer of paper approximately 1 mm in square area (image used for the front cover)

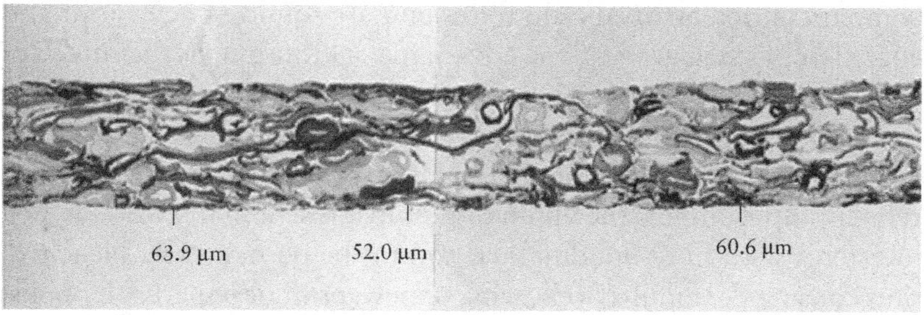

Figure 1.3 Optical microscopy cross-section of newsprint

So, here is a model to bear in mind, paper is reconstituted wood that looks like a plate of fettuccine pasta! Thinking of paper as 'reconstituted wood' is not entirely original.

Some undergraduate structural-engineering classes are tasked with a team challenge exercise to design and build a scale model bridge structure using uncooked pasta as the building material. The idea is to compete in terms of how much load the model bridge will bear before collapsing. Obviously and intuitively, the ultimate load such a model bridge will bear is a function of the structural design, but is also dependent on the properties of the integral building block, in this case, the choice of pasta. Using rigatoni (a short tube with a fluted external surface) may provide advantages over long slender fettuccine or spaghetti strands. Similarly when building churches, using cut stone blocks provides an advantage over adobe mud bricks, although the wall thickness in both cases may be over 3 m.

The properties of paper are similarly determined, to a large extent, by the fibre properties which are dependent on the wood or plant species, pulp type and any subsequent mechanical or chemical treatments. Softwood unbleached chemical pulp (kraft) fibres [7] are preferred for strength properties compared with hardwood bleached fibres [8], which are instead preferred for office copy paper. Mechanical pulp fibres, the real 'reconstituted wood', are comparably coarse in comparison to chemical pulp fibres and are found in newsprint and coated publishing grades. The advantage of kraft pulp is that much of the non-bonding lignin has been removed, allowing the cellulose fibres to bond effectively and also become irreversibly highly bleached. Mechanical pulp, traditionally groundwood or thermomechanical refiner pulp, contains the lignin originally present in the wood which interferes with fibre bonding and cannot be permanently bleached, thus causing the familiar yellowing of newsprint or paperback books.

1.1 Paper Structure, Paper Models: Fettuccine, Straw

Thinking of paper as a matrix of bonded fibres, flat wooden toothpicks, fettuccine pasta and so on, is helpful in appreciating that the properties of paper are directional. A partial alignment of pine needle straws in **Figure 1.1** can be imagined as a mattress made of straw and accordingly by analogy, paper, consisting of fibres bonded at their contact points, will be highly compressible in the out-of-plane z direction of paper (ZD). This property allows the reproduction of images onto paper using contact methods such as letterpress, offset, rotogravure and flexography, and imparts tissue tactile softness.

Most paper is made using the wet-laid process at speeds of 1,000 to 10,000 ft/min. The fibres are separated from the wood matrix and thinly suspended in water to a concentration of 0.5% or less [9]. This stock slurry is sprayed from a long slot, called the slice of the headbox, onto a wide moving straining mesh belt, historically called the 'wire', through which much of the water drains leaving a mat of wet fibres on the moving wire to be later lifted off the wire and compacted and dried to form the finished paper product. Spraying the fibre slurry at one velocity out of the paper machine headbox slice onto a draining wire moving at a different velocity, along with the finite time required for the water to drain through the wire, has two significant effects on the resulting fibre mat: orientation [10] and formation [11]. The hydrodynamics that occurs during a few milliseconds on the moving wire orientates fibres along the machine direction of machine-made paper (MD). The paper fibre mat, in various states of dewatering, drying and compacting, is pulled under tension in the MD causing stresses and further preferential orientation in the MD. The consequence is that most machine-made papers are strongest in the MD orientation, whereas the cross, orthogonal direction to the machine direction of paper (CD) is often weaker by approximately 1.5× or more. Thus, we have the principal directions of paper defined as MD, CD cross-direction and ZD shown schematically in **Figure 1.4**.

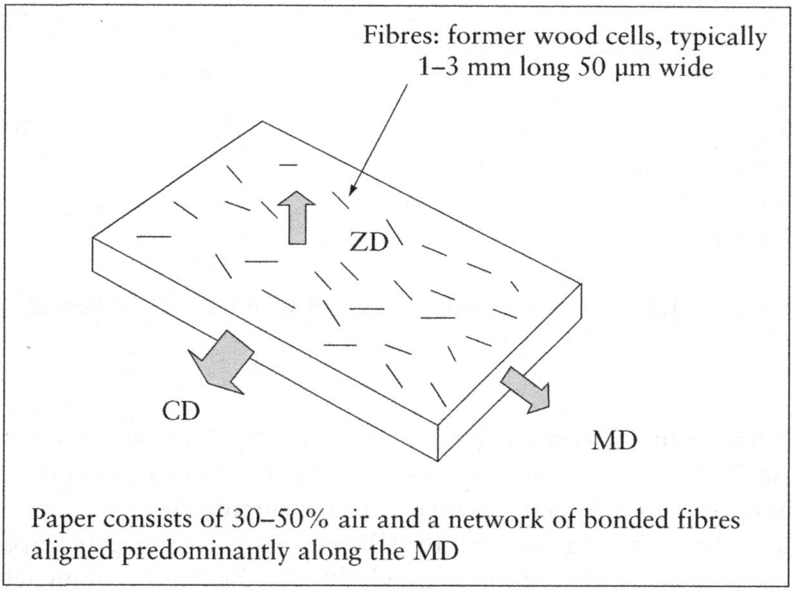

Figure 1.4 Schematic representation of paper structure showing the three principal directions of oriented machine-made paper. Fibres are depicted as dashes on the surface

This directionality in paper may be appreciated by attempting to rip rectangular pieces out of newsprint without using scissors. Most newspapers are pulled left to right in the reading direction. A paper roll will be made with the MD perpendicular to the axis roll, as shown in **Figure 1.5**.

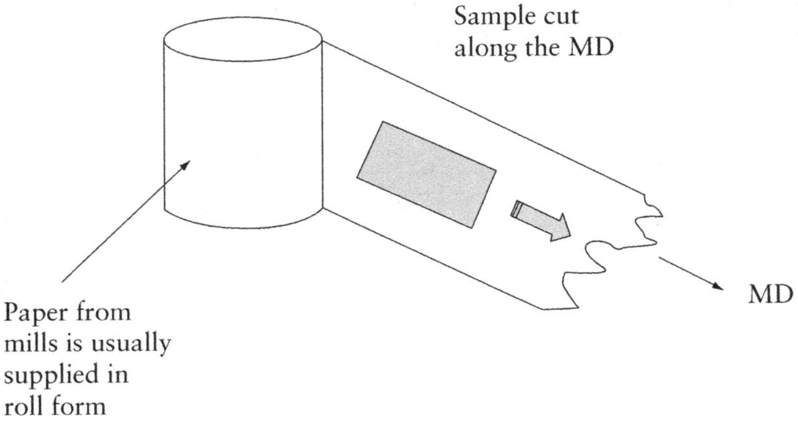

Figure 1.5 A machine-made paper roll with the MD indicated

When tearing across the page along the print direction, the tear usually follows a nice straight line along the direction of tear as this is along the fibre direction. However, when attempting to tear newsprint down or up the page which is the CD, the tear will propagate across the fibres and will not be in a straight line, leading one to ultimately reach for the scissors in order to proceed without frustration.

1.2 Formation of Paper: In-plane Lumpiness

During the paper machine former stage, the fibres in the slurry on the moving drainage wire become aligned but also tumble as the water is drained, which allows them to entrain neighbouring fibres and intertwine [12]. This leads to paper becoming non-uniform in mass on a small scale, i.e., of the order of fibre dimensions. This non-uniformity can be seen by eye by holding paper up against a light and noting the light and dark areas, the most discernible would be approximately 5 mm in size. Light areas represent lower mass than the surrounding darker areas. **Figure 1.6** is a transmitted light image of a 64 mm square of corrugated board using a night vision camera sensitive in the near infrared wavelength region, which scatters less through paper. Beside that image is a standard reflected light image of the same linerboard separated from the fluting which now shows dark iodine-stained glue lines and the agglomeration of starch from a faulty adhesive application. The transmitted image shows a mottled appearance corresponding to the non-uniform distribution of fibres in clumps that are roughly the same scale between the glue lines, approximately 8 mm in this case. These light areas, corresponding to lower local mass, will accept ink differently leading to undesirable mottle in printed images [13], and will be weaker in strength than adjacent darker areas leading to potential failure at those points when the paper is placed under tension. A considerable amount of effort in paper manufacture is focused on minimising the severity of this fibre clumping. Fast drainage speeds and low slurry consistencies are two of the simpler strategies already mentioned. Paper non-uniformity is an inevitable result of being comprised of contacting fibres, much like the clumping of straw strewn on a floor or a road as in **Figure 1.1**.

Paper consists of clumps of intertwined fibres sticking to each other *via* hydrogen bonding at their contact points. Thus, the surface is actually an ill-defined boundary which depends on the degree of fibre compaction and collapse. This out-of-plane structure complicates the measurement of paper roughness [15], caliper [16] and porosity [17]. In-plane, the non-uniformity originating from the clumping of fibres influences the test results making them affected by test specimen size, and introduces variability or a coefficient of variation (cv) of several per cent in nearly all measurements of physical properties, many of which are proportional to the basis weight. The inherent variability of paper causes considerable consternation for circumstances where paper properties are required to meet stringent quality criteria. The average value of a highly variable paper property must be set high to avoid samples that are below a set product specification. The proportionality of strength properties on basis weight means manufacturers must often meet the marketing specifications by increasing the basis weight of their products at a loss of profitability.

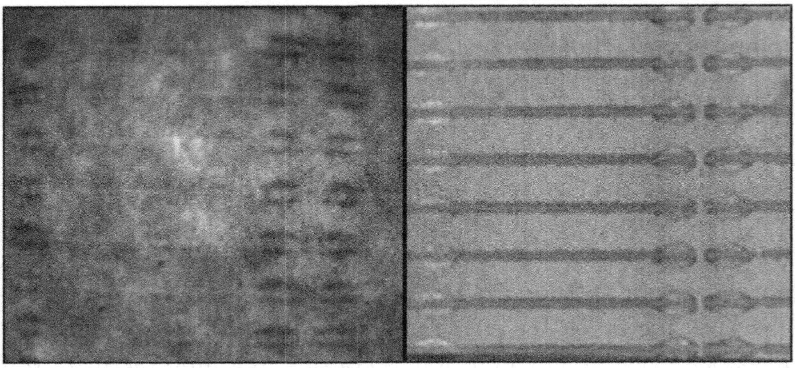

Figure 1.6 Left is a transmitted light image through corrugated board showing fluting glue lines and right is the corresponding reflected light image. Reproduced with permission from S. Johnson and R. Popil, *International Journal of Adhesives and Adhesion*, 2015, **59**, 105. ©2015, Elsevier [14]

1.3 Hydrophilicity – Paper Really Sucks – Water That Is!

The hydrophilic nature of cellulose enables paper to hold together and allows it to be recycled [18]. However, the propensity to retain moisture has consequences for paper testing, i.e., the moisture exposure history of a sample has to be considered [19]. An extreme example of the effects of moisture, which many people experience, is the accidental wetting of the pages of a book. Here, the pages of the book expand, warp out-of-plane and once dried will never be the flat smooth thin pages that they once were. In paper testing, samples of course, must be kept dry, but the ambient humidity must also be consistent and controlled. All paper contains some degree of moisture due to equilibration with ambient humidity. In high-humidity environments such as refrigerated rooms or uncontrolled warehouses of the southern US, the moisture content percentage in paper (%M) on a wet basis (i.e., weight of water/total weight as per **Equation 1.1**) can be 12% or more:

$$\%M = \frac{\text{wet} - \text{dry (weight)}}{\text{wet weight}} \times 100 \qquad (1.1)$$

In an arid environment, such as Mexico City in winter, the paper moisture will be around 4%. Some press rooms in Mexico City find that sheets of boards for printing curl up into tubes and will not run through the press. In southeastern US, stacks of boxes containing milk jugs topple over crushing the plastic milk jugs they contain. These phenomena are consequences of the hydrophilicity of paper.

Accordingly, paper is manufactured to specifications for an environment at 50% relative humidity (RH) and room temperature (RT). The moisture of paper in these conditions is usually approximately 7.5%. Rolls and reams of paper products are wrapped in moisture-proof covers during shipping so that the paper will have the expected specifications at 50% RH. Thus, to be consistent, paper testing is performed at 50% RH with an RT of 23 °C [20]. If testing were to be performed at a lower humidity, the paper moisture would become lower and many strength properties would become proportionally higher. Conversely, if paper were to be tested at a higher humidity than 50%, the paper moisture would become higher and the strength properties would be proportionally lower from those specified at 50% RH. **Figure 1.7** shows a moisture history curve of a paper's moisture content adjusting to relative ambient humidity.

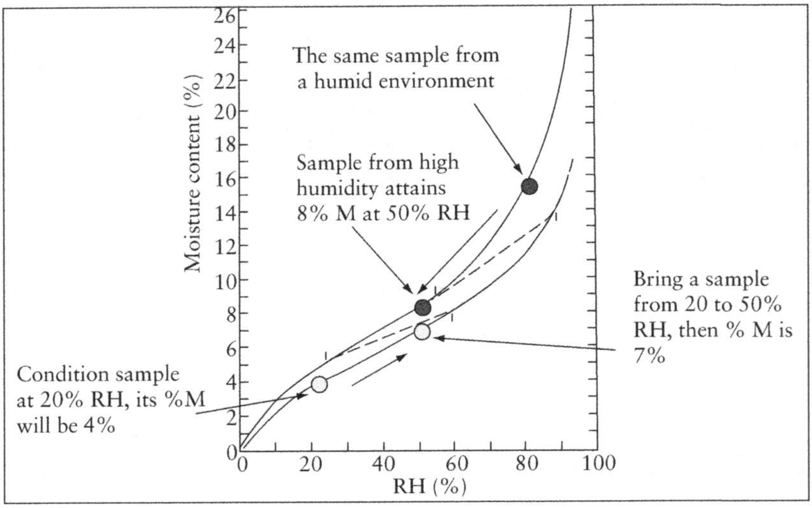

Figure 1.7 A hypothetical paper moisture curve showing the history cycle of increasing and decreasing ambient humidity

If paper is left to equilibrate to a high-humidity environment (>50%), as happens with unwrapped rolls of paper left in an uncontrolled press room or warehouse for long periods of time, and tested at 50% RH, the strength properties will be measurably lower than when originally measured. This is another consequence of the hydrophilic nature of paper and can be described as the moisture hysteresis of paper. When paper is exposed to high humidity and equilibrated to a lower humidity, it will have a slightly higher humidity than paper brought from a lower humidity to the testing standard humidity of 50%. The aim for a paper testing laboratory is to obtain consistent reproducible property values. Incoming samples often have an unknown moisture history so the standard practice is to equilibrate paper samples to a low-moisture state followed by equilibration to the 50% RH and perform the testing at 50% RH.

The question often arises as to what the duration of preconditioning/conditioning periods should be prior to testing. In production mills, paper quality is required to be verified quickly when machine or stock changes are made. The Technical Association of the Pulp and Paper Industry (TAPPI) method protocol of 24 h sample storage at 20% RH followed by 24 h at 50% RH is intended to encompass reams of paper sheets and boards. However, a single sheet suspended in moderately moving air will equilibrate to 90% of its final moisture value in 15 min and fully in 2 h. Some mill testing labs have adopted the practice of placing samples into a 1 kW microwave oven for 10 s to ensure the paper is dried and then suspending the sheets in 50% RH moving air for 2 h prior to testing strength properties.

Figure 1.8 shows the exponential change in time of a moisture-dependent property of dry paper acclimating from a dry state to its equilibrium point in a humid ambient environment. Of course a longer time is better, but a minimum can be 1 h for a single sheet in circulating air flow or 5 h for a corrugated board [21]. Nonetheless, urgency requires rapid measurements in a mill production situation and measurements are taken of samples at various moistures. In such cases, a corrective factor may be applied if the moisture content of the sample is known at the time of measurement, as product specifications are based on paper equilibrated to 50% RH. One study, requested by a mill for softwood kraft linerboard, required obtaining a moisture correction for measurements of compression strength, which is important for corrugated box manufacture. The specific compression strength of interest here is called the short-span compression test or strength (SCT). Measurements for a moisture corrective factor were made for the mill samples equilibrated in various humidity values from 20 to 80% and the results are shown in **Figure 1.9**.

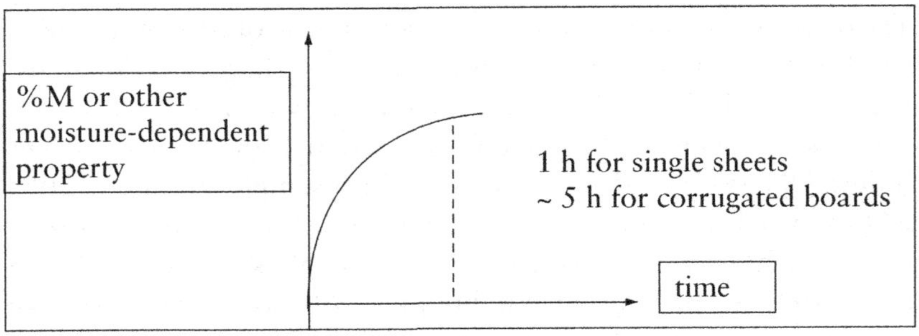

Figure 1.8 Hypothetical %M, strength, basis weight *versus* time for dry paper equilibrating in a humid environment

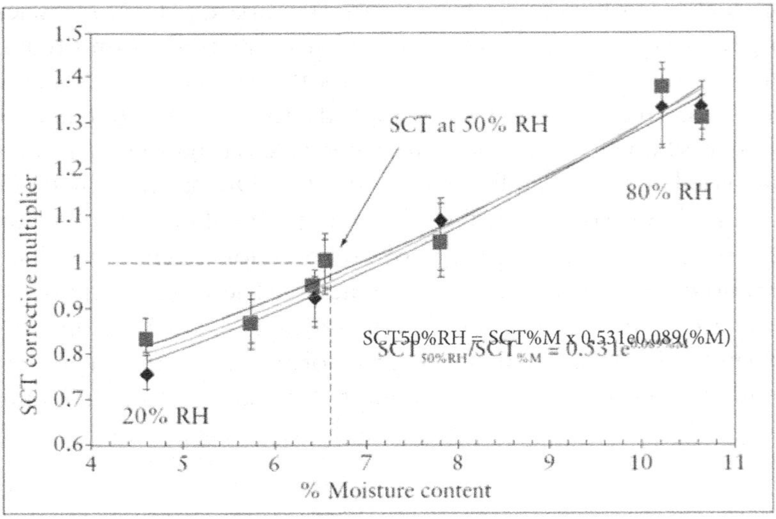

Figure 1.9 SCT of kraft linerboard measured at various humidities relative to the standard value of 50% RH

Basic theory predicts that the dependence of strength properties on moisture will follow exponential behaviour, and the results are expressed as a corrective factor to the values at 50% RH in **Equation 1.2.** An increased paper moisture content has the effect of decreasing the fibre elastic modulus and accordingly, strength values follow the same trend.

$$SCT_{50\%RH} = SCT_{\%M} \times 0.531e^{0.089(\%M)} \tag{1.2}$$

To a very good approximation, followed by many instruments that correct for moisture, a linear interpolation of the strength moisture dependence is used and shown in **Equation 1.3**:

$$strength_{50\%\ RH} = strength_{(\%M)} \times [(\%M \times)0.07 + 0.47] \tag{1.3}$$

1.4 Paper Property Variability – Why Paper Testing is Necessary

All paper has variability on a millimetre scale due to it being composed of bonded fibres, as discussed above. The manufacturing process introduces yet another variability on the scale of the test specimen size of several centimetres. Paper is often manufactured in high volumes on a high-speed paper machine which produces at its end, a jumbo roll of paper on the reel being some 20 tonnes perhaps 20 to 60 ft wide, a schematic representation is shown in **Figure 1.10.** There are inevitable profiles in the CD [22] that arise from stock velocity differences, which fall to zero at the edges of the forming wire, drying temperature differences across the machine and lack of tensile restraint in the CD.

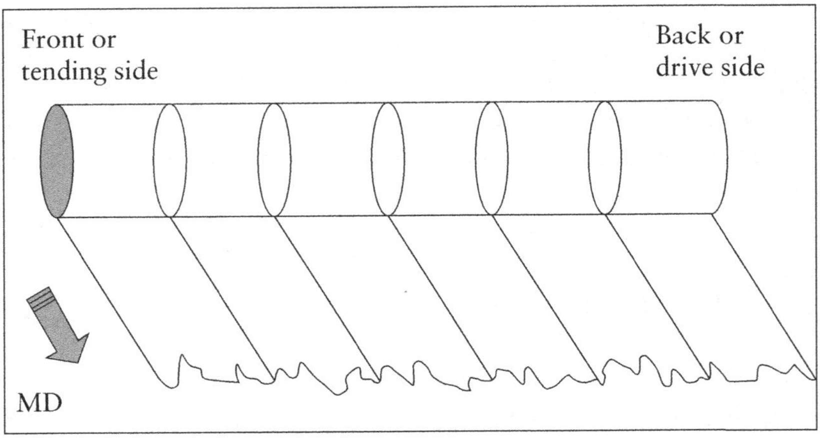

Figure 1.10 Schematic representation of the jumbo roll at the dry end of a paper machine

Paper machine rolls used for pressing and calendering introduce a pressure profile that may not be compensated by the eventual wear of the roll or the supporting bearings. Along the MD, pulsation in the pressures at the stack headbox will manifest as variation in basis weight, any press or calendering rolls that are not perfectly round will also introduce a basis weight variation. Press felts with any defects, forming wire anomalies and uneven seams, introduce periodic artefacts in the sheet [23]. On a longer time scale, there are also variations in the raw material of the fibre stock furnish arising from both variations in the stock preparation process and the supply of the raw material, wood or recycled materials. The combination of all these variations requires any testing of paper to consist of repeated tests on the replicates of the paper sample to account for the inherent variability. Typically, 10 repeats are made for each sample when testing for tensile or compression strength. Repeat measurements are used to examine the result such that if the cv exceeds a typical value of 7% or so, then either the sampling method, sample preparation or the instrument becomes suspect and requires investigation.

1.5 Will it be Paper or Plastic? – Why Ask?

Typical variability in any paper can be expressed as the cv, (standard deviation/average value) of whatever property is of interest. The clumping of fibres from the wet-laid straining process leads to a mass variation of around 6 to 8% [24]. Strength is usually proportional to paper mass so it follows that paper strength properties also display a cv of this order. This variability is what governed the retailers' answer to whether paper or plastic would be used to bag groceries, i.e., usually plastic. A shopping bag is made to meet strength specifications but to reduce basis weight it will not have a strength value much above the specified lower limit. The variation in strength will mean there will be weak spots in the paper bag where a tear may initiate. Paper under tension will not stretch very much, the fibres become pulled from one another and a tear opens up at a weak low basis weight point which then propagates along similar low basis weight points, resulting in groceries falling onto the pavement. Plastic bags do not have the cv% variability of paper and the stretch to failure exceeds that of paper by approximately 400×, allowing one to realise long before failure that a bag is about to fail.

1.6 There are Two Sides to Every Sheet of Paper

Paper sheets have two sides that usually appear to be the same but are always different to some extent. The wet-laid stock straining process of forming paper imposes asymmetry in the cross ZD profile. Traditionally, the two sides of paper refer to the fourdrinier paper machine, where sheets are formed by draining water *via* gravity on the moving straining wire mesh and the resulting mat is pressed free from water on the top by a moving belt of an absorbing felt fabric. Accordingly, fourdrinier-made papers are said to have a 'top' or 'felt' side and a 'bottom' or 'wire' side. The direction of the free water being drained from the fibre mat is towards the wire side. The stock consists of fibres which inevitably have a length distribution such that there is usually a larger proportion of shorter fibres and submillimetre fibre fragments called 'fines'. The directional drainage of the stock combined with the interstices of the forming wire results in the loss of the fines stock fraction at the wire side. Moreover, contact of the mat with the wire will superimpose the topography of the wire fabric onto the bottom side of the sheet.

The consequence is that the top or felt side of the sheet will often be smoother, the surface having fewer voids, and the bottom or wire side will be rougher with more voids between the long fibres. The structure results in the two sides of the sheet absorbing water differently. Printing and writing will appear differently on the two sides of the sheet [25], especially when using aqueous inks. Applying a coating will also have different results on a fourdrinier sheet.

Air flow is somewhat affected by the two sidedness, much like forced air-furnace filters, papers are more permeable when air flow is towards the more open porous wire side. The in-plane ductility of the sheet is also affected by two sidedness. Fibres in the wire side will be surrounded with less bonded fines, which act as a filling cement to transfer stresses so that when in tension, the wire side surface often may have lower strength. This is consequential whenever paper is stressed in a curved geometry, such as in the burst test where paper is punctured while being secured over an expanding rubber diaphragm or, when paper is folded or scored when making carton packaging. When folding paper, it is preferred to have the felt side on the outside of the fold so it can endure the higher stresses from the imposed curvature and thereby minimise the likelihood of cracking.

1.7 Twin Formers Nomenclature

Obviously, two sidedness of paper is not a desirable property, so considerable effort has focused on forming technology to minimise or eliminate the asymmetry. There are twin-wire formers, roll formers or top former retrofits. The idea here is to dewater from both sides of the sheet as simultaneously as possible; nonetheless, some two sidedness remains. The paper is carried to the press section by the conveying wire side of the paper (CW) while the opposite side of the sheet is in contact with the backing wire (BW) side of paper.

Whenever water is drained from a consolidating fibre mat, the fibre fines and filler-particulate matter are carried with the water so that in the case of twin-wire forming, the middle of the cross-section of the sheet is proportionally devoid of matter compared with the CW or BW surfaces. The lower density in the middle of the sheet on twin-wire formed papers makes the paper weaker in the middle so that if the sheet is split by applying tension to both sides of the sheet, as done in some types of paper tests, the twin-wire formed sheet will split evenly through its thickness. In contrast, fourdrinier-made papers will split out-of-plane proportionally weighted to the felt side since the felt side has retained more of the fines which serve to bond themselves and the fibres to each other.

1.8 The Many Types of Paper

The term 'paper' here will apply to all wood fibre-based materials formed by the wet-laid fibre stock draining process. Most people are familiar with the common office copier paper, which is made of bleached kraft chemical pulp. This type of paper is usually fourdrinier-made, of medium basis weight or weight per unit area, and is comparatively porous and rough, which is of no consequence for non-contact printing or writing with viscous inks. Papers encountered in journals will often be comparatively lightweight, coated and compressed to a higher density to allow a high-gloss surface and good reproduction of images. To reduce costs, the base of such papers will be mechanical pulp, traditionally groundwood but more likely thermomechanical pulp mixed in with some kraft. Bleaching of pulp is chemically expensive and weakens the strength of cellulose fibres, so corrugated boxes are made from unbleached kraft of high basis weight, as strength is proportional to basis weight. Paper roll cores are tubes made of many bonded layers of heavyweight unbleached kraft. Similar heavyweight papers are cartonboards for packaging consumer goods that are often coated on one side to allow colour printing.

On the opposite scale to the heavy basis weight of packaging and printing papers are the tissue and towel products. Here, the combination of absorbency and softness with strength require low-density bleached kraft. Towel papers contain wet strength polymeric chemicals, whereas sanitary tissue may contain lubricants to produce more tactile softness. The demand for these products to have a bright white appearance requires bleached chemical kraft pulp, and the tactile softness marketing specification requires the fibre species to be predominantly hardwood, often eucalyptus from South America.

1.9 Summary

Describing paper as ‘reconstituted wood’ accounts for wood fibre wet-laid paper being comprised essentially of self-bonding matrices of fibres. The fibres are usually much longer, with lengths extending to millimetres, than wider and are flat in cross-section. Papers have a wide range of densities, with higher densities required for writing printing and low densities are required for absorbency. Mechanical pulps produce weaker less bright papers such as newsprint, whereas chemical pulps produce weaker but often bleached and white papers. Containerboard corrugated packaging is predominantly made with unbleached chemical pulp for optimal strength per basis weight. Machine-made paper has an orientation where fibres aligned in the MD provide increased strength compared with cross-direction CD. Fibres lying flat in-plane provide a high compressibility in the ZD. Paper fibres, being comprised of the polysaccharide polymers cellulose and hemicellulose, are hydrophilic and retain moisture absorbed from the ambient atmosphere. The amount of moisture in the paper proportionally affects its strength properties, thus care must be taken to ensure paper samples have the desired moisture level prior to testing.

References

1. D. Hunter in *Papermaking: The History and Technique of an Ancient Craft*, Dover Publications, Inc., Mineola, NY, USA, 1978.

2. G. Smook in *Handbook for Pulp and Paper Technology*, Angus Wilde Publications, Vancouver, Canada, 1992.

3. C. Ververis, K. Georghiou, N. Christodoulakis, P. Santas and R. Santas, *Industrial Crops and Products*, 2004, **19**, 3, 245.

4. U. Hirn and R. Schennach, *Scientific Reports*, 2015, 5, 10503.

5. R.C. Howard and W. Bichard in *MRS Proceedings*, Cambridge University Press, Cambridge, UK, 1992, **266**, 195.

6. S.M. Potter, *Current Biology*, 1996, **6**, 12, 1595.

7. R.A. Horn in *Morphology of Wood Pulp Fiber from Softwoods and Influence on Paper Strength*, FSRP-FPL-242, Forest Products Lab, Madison, WI, USA, 1974.

8. R.A. Horn in *Morphology of Pulp Fiber from Hardwoods and Influence on Paper Strength*, FSRP-FPL-312, Forest Products Lab, Madison, WI, USA, 1978.

9. C.J. Biermann in *Handbook of Pulping and Papermaking*, Academic Press, San Diego, CA, USA, 1996.

10. T.R. Hess and P.H. Brodeur, *Journal of Pulp and Paper Science*, 1996, **22**, 5, J160.

11. D.S. Keller and P. Luner, *Review of Scientific Instruments*, 1998, **69**, 6, 2495.

12. R.J. Kerekes, *Nordic Pulp and Paper Research Journal*, 2006, **21**, 5, 98.

13. J-P. Bernie, H. Pande and R. Gratton, *TAPPI Journal*, 2006, **5**, 10, 28.

14. S. Johnson and R. Popil, *International Journal of Adhesives and Adhesion*, 2015, **59**, 105.

15. R. Xu, P.D. Fleming, A. Pekarovicova and V. Bliznyuk, *Journal of Imaging Science and Technology*, 2005, **49**, 6, 660.

16. W.A. Wink and G.A. Baum, *TAPPI Journal*, 1983, **66**, 9, 131.

17. C. Hii, Ø.W. Gregersen, G. Chinga-Carrasco and Ø. Eriksen, *Nordic Pulp and Paper, Research Journal*, 2012, **27**, 2, 388.

18. M.A. Hubbe, R.A. Venditti and O.J. Rojas, *BioResources*, 2007, **2**, 4, 739.

19. W.A. Wink, *TAPPI Journal*, 1961, **44**, 6, 171.

20. TAPPI T 402-om-88: Standard Conditioning and Testing Atmospheres for Paper, Board, Pulp Hand Sheets and Related Products, 1992.

21. S. Allaoui, Z. Aboura and M.L. Benzeggagh, *Composites Science and Technology*, 2009, **69**, 104.

22. S.M. Hoole, S.J. L'anson, M. Ora, T.N. Ashworth, D. Briggs, B. Phillips and R.W. Hoyland, *Paper Technology*, 1999, **40**, 10, 63.

23. S. McLeod, Z. Nesic, M.S. Davies, G.A. Dumont, F. Lee, E. Lofkrantz and I. Shaw in the *Proceedings of the Dynamic Modeling Control Applications for Industry Workshop*, IEEE Industry Applications Society, 30th April–1st May, Vancouver, Canada, 1998, p.59.

24. J.M. Hellawell, *Paper Technology and Industry*, 1973, **14**, 1, 24.

25. I.I. Pikulik and J.D. McDonald, *TAPPI Journal*, 1987, **70**, 4, 75.

2 Tensile Properties

The most fundamental and informative physical evaluation for paper is probably that of the tensile test. It provides the elastic properties of paper which are directly related to fibre quality and the level of fibre bonding is most aptly described in the 1982 paper by Page and Seth [1]. To a large extent, most of what is required to be known about a paper sample can be gleaned from a tensile test provided it is interpreted correctly. Therefore, tensile testing is routinely used in manufacturing operations for quality control.

Simply stated, the tensile test is a strip of paper clamped at both ends and pulled at a constant rate until failure [2]. The load at failure is the tensile strength (S_t), which is satisfactory for many purposes but much more can be gleaned from consideration of what happens while the strip is pulled to fracture.

The tensile test pulls paper either in the machine direction of machine-made paper (MD) or the cross direction of paper (CD), so strips are cut such that the length is along either the MD or CD. As described in **Chapter 1**, the MD is the length direction of paper pulled from a roll. The fibres comprising the paper are aligned in the MD and so strength properties are generally higher in the MD than the transverse CD. The strip is firmly clamped at either end leaving a free gap of several inches in length. A load cell is attached to one of the two clamps and is affixed to a frame. A typical universal test frame fitted with clamps for tensile testing is shown in **Figure 2.1**.

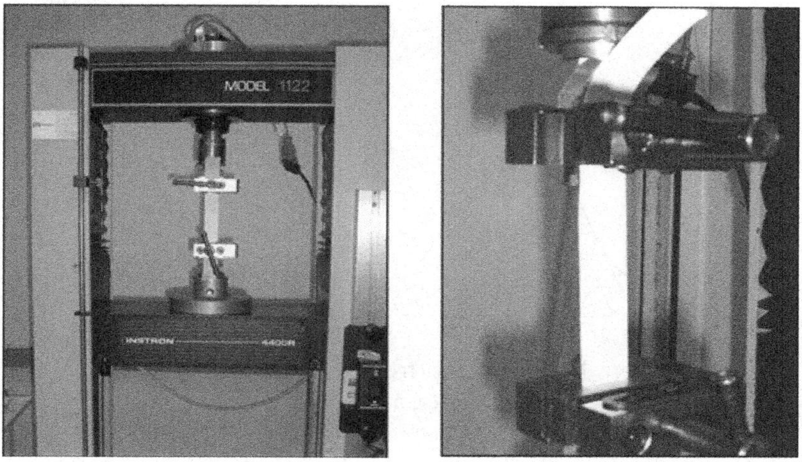

Figure 2.1 Photograph of a test frame for tensile testing, left-hand side shows a fixed load cell clamp, test strip and another clamp attached to a moving cross-head, and right-hand side shows close-up detail

In this case, the test frame is connected to software which controls the test and collects the load cell and displacement data for analysis. The cross-head is connected to motor-driven screws and the rate of downward motion is typically set to 25 mm/min. The motion of the cross-head increases the tension load 'F' along with increasing the displacement along the strip length of 'Δl'. A graph of tension load 'F' *versus* time for typical copy paper would look like **Figure 2.2**.

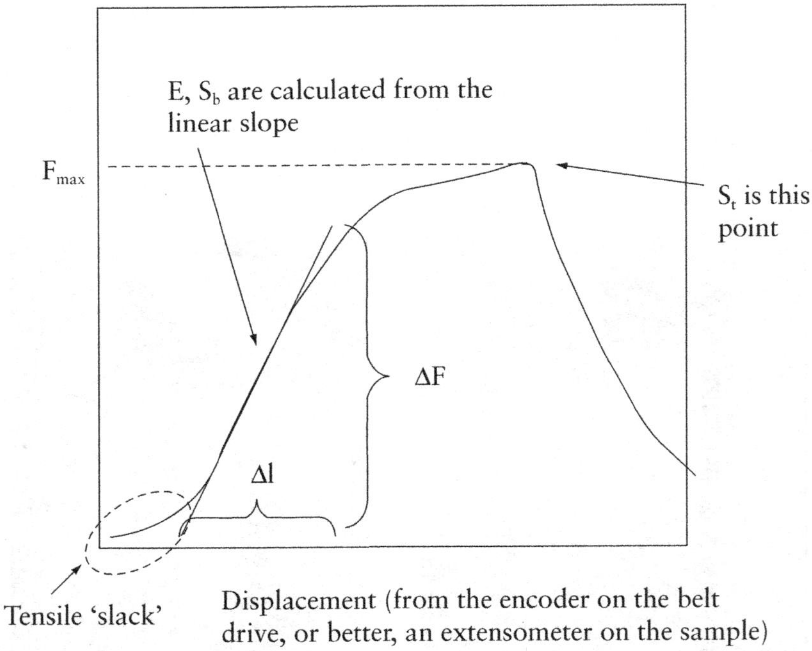

Figure 2.2 A hypothetical load-displacement curve for a paper tensile test

2.1 Basic Mechanics of the Tensile Test.

Once slack is removed from the sample, if it is mounted loosely, the load increases steadily in what is called the elastic region. The rise in the load, 'ΔF', in typical units of newtons 'N', and the change in displacement Δl of a strip with free length 'l' divided by the width 'w' of the test sample provide a useful quantity called the tensile stiffness 'S_b' defined by **Equation 2.1:**

$$S_b = \frac{\Delta F \times l}{\Delta l \times w} \tag{2.1}$$

N/mm or equivalently kN/m are the units for S_b. Strength for paper is also cited per unit width so also carries force per unit width N/mm. The S_b is defined as the elastic modulus (E) multiplied by the effective caliper 't' in **Equation 2.2.** The modulus is the material resistance to mechanical deformation given by Hooke's law and for paper and many other materials is directionally dependent [3], so a strip cut along the MD is ascribed a MD suffix or conversely a CD suffix for the case of a CD test.

$$E_{MD,CD} \times t = S_{b_{MD,CD}} \tag{2.2}$$

Strain is the relative displacement 'Δl/l' denoted in mechanics as 'ε' and stress, the force per unit area, is denoted 'σ'. In the case of the uniaxial tensile test strain defined by **Equation 2.3:**

$$\varepsilon_{MD,CD} = \frac{\Delta l_{MD,CD}}{l_{MD,CD}} \tag{2.3}$$

Here for simplicity and as an approximation, we are ignoring the small Poisson effects, i.e., when a tensile load is applied on a paper strip cut along one direction, there will also be some dimensional changes in the orthogonal and out-of-plane directions, which can be described by a directional Poisson ratio 'ν'. Similarly, for stress applied to a paper strip we have **Equation 2.4:**

$$\sigma_{MD,CD} = \frac{\Delta F_{MD,CD}}{t \times w} \tag{2.4}$$

So, the S_b can be cast into the familiar form of Hooke's law in **Equation 2.5** as:

$$E_{MD,CD} = \frac{S_{b_{MD,CD}}}{t} = \frac{\sigma_{MD,CD}}{\varepsilon_{MD,CD}} \tag{2.5}$$

E appears in formulas for other mechanical properties such as bending stiffness and the speed of sound in paper, which will be explored later in this book. E is dominated by fibre quality in most papers and can be used to predict what the paper strength will be. The 't' in the above equation is the 'apparent' thickness of paper, but as stated previously, the surface of paper is rather ill defined due to the combination of surface unevenness and paper compressibility, however this issue will be resolved in a later chapter.

A yield point is reached where departure from linearity occurs in the tensile load displacement curve as the load increases further with increasing displacement. Paper stretched beyond this point is said to be in non-recoverable plastic deformation. Metals stretched beyond this point are said to be strain hardened. If tensile load is relaxed beyond this yield point, a permanent deformation results. If tension is increased beyond the yield point, the displacement increases until the sample strip fractures and the tensile load drops to zero. The load where the strip fractures is called the ultimate strength or strength of the paper. There is usually a relationship between the strength and the linear reversible S_b values; however, strength can be altered somewhat through the inclusion of additives in the pulp stock leaving the S_b unchanged.

The work 'W' done on the sample during the tensile test is the area covered by the load displacement curve defined by **Equation 2.6**:

$$W = \int_0^{l_f} F\,dl \tag{2.6}$$

where 'l_f' is the displacement or elongation at failure, and 'dl' is the incremental displacement.

This is useful to describe the elastomericity of the sample and is a combination of the sample failure strength and stretch or strain to failure. Samples consisting of long, highly curled fibres would have a high strain to failure, and so a higher W, than samples comprised of short straight fibres. Although the strain to failure $(l_f - l_0)/l_0$ is usually larger in the CD (~4%) than the MD (~1.5%) for most machine-made papers, W is lower for the CD due to the lower strength. The tensile work of a sample is sometimes interpreted as a measure of sample resistance to in-plane tear and may be useful for a comparative ranking of samples for potential tear resistance. However, measurements of fracture toughness, which are tensile tests of specifically nicked samples [4], are more accurate for this purpose. The TAPPI method for reporting the tensile work or, equivalently, the energy absorbed (tensile energy absorbed) is normalised to the transverse area under tensile load, i.e., area of the test strip between the clamps and so the units are J/m^2.

2.2 Effect of Test Specimen Size on Tensile Strength

Since paper consists of non-uniform clumps of fibres, the size of the sample test piece has an effect on the results [5]. The length direction of the sample strip aligned along the MD or CD, as accurately as possible, should be several times the width. When the length of a tensile test span between clamps is shortened to less than 1 mm, the fibre network effects become diminished and the results are considered to be dominated by the fibre S_t [6]. Indeed, zero span tests are used as a quality check of fibre quality [7] and interpreted using the 'Page equation' model for S_t [8].

Dimensions of the strip for paper tensile testing should be several times the length of the longest fibres. Intuitively, strength can be understood to become lower with longer strip specimens, as paper is non-uniform in mass distribution due to the clumping of fibres. Therefore, the larger the test piece, the greater becomes the likelihood of stressing a weak part in the test piece.

Two sizes are commonly used: 10 × 1" and 160 × 15 mm. The latter is more popular as it is also a convenient size when testing laboratory-made handsheets and the 15 mm width strips are also used for short-span compression, hygroexpansivity testing.

Paper fibres are composed of polymeric materials that exhibit viscoelastic properties, which means that the rate of mechanical deformation also has an effect on the results of a test. Generally, strength properties increase approximately 7.5% for every decade change in timescale [9, 10]. Paper will appear to be stronger if it becomes stressed at a faster rate. In most paper tests, two common deformation rates, probably originating from a historical available equipment convenience, are either 0.5 or 1 inch/min.

The TAPPI method T 494 contains a footnote that if a 1" wide sample is used with a 7" span length, the displacement rate should be 1 inch/min, but if the 160 × 15 mm sample is used instead, the displacement rate should be 0.5 inch/min [2].

2.3 A Study of the Effect of Sample Size and Deformation Rate

What happens when we use two different rates and two different sample sizes in the tensile test? Are the differences significant? Changing the rate from 0.5 to 1 inch/min significantly speeds up the testing time, especially since the usual protocol stipulates that 10 replicate tests be made for a sample. Are 10 replicate tests necessary, would 20 be better or would 5 suffice?

Let's examine the possibilities. As representative paper samples for this testing exercise, we will take a typical unbleached kraft linerboard of 205 g/m^2, bleached kraft copier paper of 75 g/m^2, lined kraft writing paper of 56 g/m^2 and newsprint of 49 g/m^2. A comparison will be made between sample test piece specimen size and displacement rates. The matrix of experiments is listed in **Table 2.1**.

Table 2.1 Matrix of test piece size and deformation rates for the tensile test	
Test Piece Size	**Displacement rate (inch/min)**
7 × 1”	1
7 × 1”	0.5
160 × 15 mm	1
160 × 15 mm	0.5

Results for a typical lightweight lined bleached white notebook paper and a typical medium weight unbleached kraft linerboard are shown in **Figure 2.3** for the MD and **Figure 2.4** for the CD. To compare results for the different basis weight samples, the strength value in N/mm results are divided by the sample basis weight so that the results are presented as indices, hence the odd looking mixed units of N-m^2/g-mm. Tensile properties scale linearly with basis weight, so dividing value by the basis weight is a standard practice to compare samples of differing basis weights. For the case of the notebook paper, the largest difference occurs when using the larger 1” wide strip compared with the 15 mm strip.

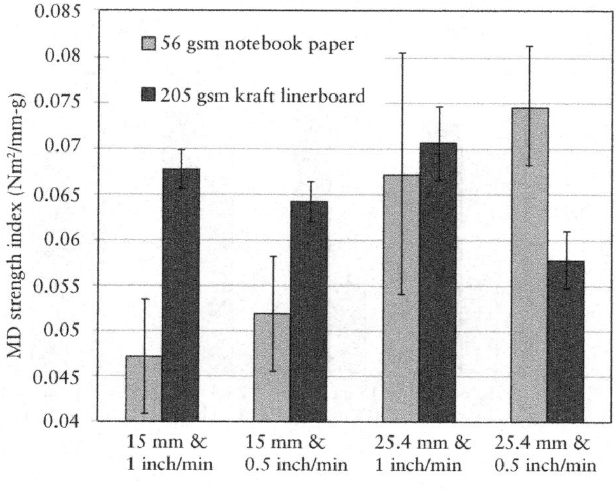

a)

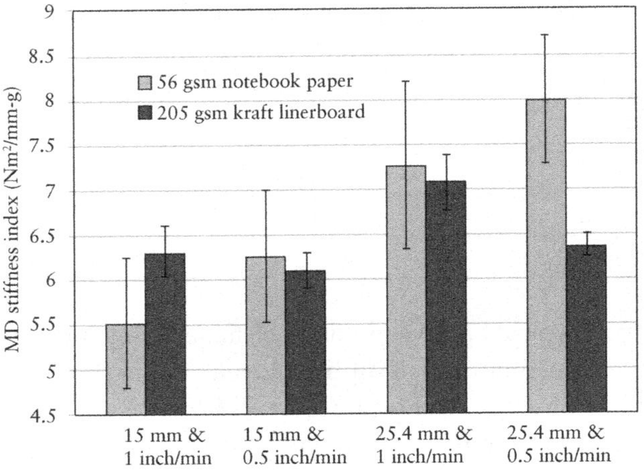

b)

Figure 2.3 Tensile test results for MD strength (a) and stiffness (b) for notebook and kraft linerboard papers using two different strip widths and elongation speeds

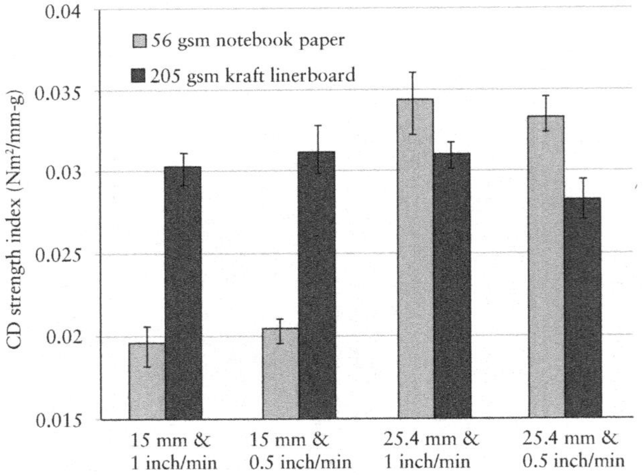

a)

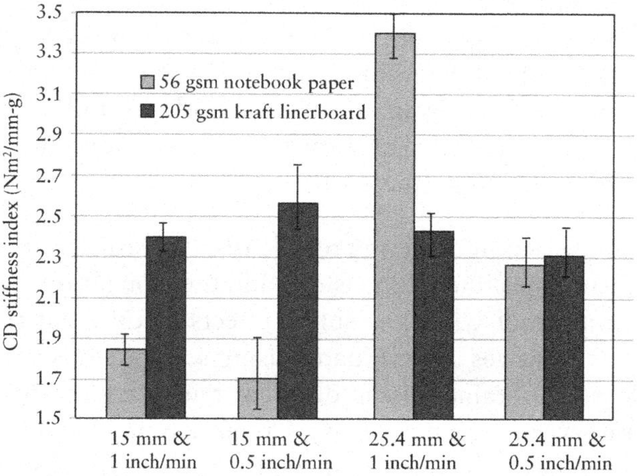

b)

Figure 2.4 Tensile test results for CD strength (a) and stiffness (b) for notebook and kraft linerboard paper using two different strip widths and elongation speeds

S_b are also lower when testing a 15 mm strip compared with the 25 mm width, approximately 23% lower in the MD and 37% in the CD. Thus, testing a smaller width results in lower strength and stiffness values. This may be expected on the basis of paper non-uniformity commonly called formation, which is most easily visualised when looking through a sheet of paper backlit by a bright light, although radiographs are less prone to scattering effects from fillers [11] and are more accurate representations of the mass distribution in paper. However, the mottled appearance of typical paper illuminated from the back is due to the clumping of fibres during the wet-laid process and is described as a crowding factor [12], which relates to the effect of longer fibres entangling others forming clumps in suspension then later forming the flocs we see when the suspension is drained. So floc size is dependent upon the longer fibres in suspension and are typically approximately 4 to 5 mm in diameter. The 15 mm strip width therefore may not be wide enough to cover several flocs across its width and so produces a lower S_t value. The effect of change in elongation rate from 0.5 to 1 inch/min appears not to be as consistently significant for both 15 and 25 mm wide strips. The expectation is that a lower elongation speed would result in lower strength and stiffness values.

In summary, when interpreting tensile results, care should be taken to note the test specimen dimensions and the elongation speed. This becomes important when the subtle effects of different treatments or chemical additives are evaluated using separate sets of data that may have been obtained using different tensile testing parameters. For laboratory convenience, testing 15 mm samples at the elongation speed of 25 mm/min is convenient and the results are not significantly different when using a 15 mm wide strip tested at the standard 12 mm/min speed.

2.4 Units, Breaking Length

Paper strength is always specified per unit width, e.g., N/mm. As mentioned before, it is common practice to normalise S_t and S_b values by the basis weight 'β' to account for the linear increase of S_t with basis weight. This way, the effects of fibre furnish or species changes, strength additives or filler content can be assessed independently of basis weight. Another common equivalent method is to also use breaking length 'BL', which is the calculation of the length of paper that would cause it to break under its own weight. Since the strength of force to failure 'F' in this case is given by Newton's law, F = mg, we have breaking length defined as:

$$BL = \frac{F/w}{g\beta} = \frac{\text{tensile strengh (N)}}{\text{sample width (m)} \times \text{basis weight}\left(\frac{\text{kg}}{\text{m}^2}\right) \times 9.8\,\frac{\text{m}}{\text{s}^2}} \tag{2.7}$$

which using the usual units of S_t, 'F' in N/mm, basis weight in g/m^2 becomes:

$$BL(km) = 102 \times \frac{F(\frac{N}{mm})}{\beta(\frac{g}{m^2})} \tag{2.8}$$

2.5 Summary

S_t is perhaps the simplest yet most fundamental of the physical tests for paper. A tensile test produces the strength stiffness and work values for a sample. Strength and stiffness relate to the *E* of the sheet, which is principally governed by the fibre quality for a given basis weight and sheet density. Although the test method is comparatively devoid of testing artefact and inherent high variability associated with any other tests, tensile results are dependent on the test parameters such as sample size, sample orientation and elongation rate. The capability of obtaining the S_b and in turn the *E* for the two principal directions, MD and CD, allow the inference of other mechanical properties of paper.

References

1. R.S. Seth and D.H. Page in *The Role of Fundamental Research in Paper Making*, Mechanical Engineering Publication, London, UK, 1983, p.421.

2. TAPPI T 494: Tensile properties of paper and paperboard (using constant rate of elongation apparatus), 2016.

3. R.W. Perkins in *Handbook of Physical Testing of Paper*, Eds., R.E. Mark and J. Borch, CRC Press, Boca Raton, FL, USA. 2001, 1, 2.

4. P. Makela and C. Fellers, *Nordic Pulp and Paper Research Journal*, 2012, **28**, 2, 352.
5. A. Hagman and M. Nygards, *Nordic Pulp and Paper Research Journal*, 2012, **28**, 2, 295.
6. TAPPI T 231: Zero span breaking strength of pulp (dry zero-span tensile), 2016.
7. W.J. Batchelor, B.S. Westerlind, R. Hagglund and P. Gradin, *TAPPI Journal*, 2006, **5**, 10, 3.
8. D.H. Page, *TAPPI Journal*, 1969, **52**, 4, 674.
9. R.C. Moody and J.W. Koning, USDA Research Note FPL-1212, April, 1966.
10. R.E. Popil, *BioResources Journal*, 2012, **7**, 2, 2553.
11. D.S. Keller and J.J. Pawlak, *Journal of Pulp and Paper Science*, 2001, **27**, 4, 117.
12. R.J. Kerekes, and C.J. Schell, *Journal of Pulp and Paper Science,* 1992, **18**, 1, J32.

3 Ultrasonic Testing of Paper

3.1 Introduction

One of the most useful tests for detecting the changes in relative sheet quality is sonic propagation through and along sheets. Rediscovered for paper applications in the 1960s [1], the technique is often underutilised and overlooked for paper characterisation in favour of mechanical measurements. A good review of ultrasonic measurements and applications for paper can be found in Waterhouse [2]. A major selling point for ultrasonic measurements is that it is comparatively quick, requires no specifically prepared sample test piece size and is non-destructive. These features are very useful in a testing laboratory, allowing first pass screenings of sample sets which can be followed by standard mechanical testing to ascertain any differences that may be of interest to investigate such as the effects of various pulp treatments or stock additives to enhance properties.

Many mechanical properties of a paper sheet are affected by the paper elastic modulus 'E'. The discussion regarding the mechanics of paper as an orthotropic solid can be extensive, however [3], for the purposes of quality testing related to mechanical properties of interest, the immediate concern here will be sonic propagation in the three principal directions of machine-made paper: the machine direction of machine-made paper (MD), the machine direction of paper (CD) and the out-of-plane z direction of paper (ZD). The main principle that is useful for the sonic testing of paper is the relationship between paper modulus and the speed of sound, 'V', as longitudinal waves can be simply given by the approximation in **Equation 3.1:**

$$E \cong \rho V^2 \tag{3.1}$$

where the 'apparent density' of the paper test sheet is the basis weight divided by its caliper, preferable and more accurately, it is the 'soft-platen' caliper, for reasons that will be explained shortly.

The modulus relationship in **Equation 3.1** becomes approximately 10% more accurate with the inclusion of Poisson constant ratio terms, which are calculated [4] using the measurements of in-plane shear sonic wave propagation and results from orthotropic solid mechanics. However, in the author's experience, the added complexity in doing so does not provide a useful advantage.

As sonic propagation in paper is directionally dependent, as is the modulus, both 'E' and 'V' are written with suffixes 11, 22 or 33 corresponding to MD, CD and ZD, respectively. Typically, for many papers commonly encountered, 'E' and 'V' are smaller in the CD compared with MD by approximately 1.5 times or more, and ZD values are approximately 30 times or more lower than either MD or CD values. The ratio of MD to CD values is attributable to the combination of fibre orientation and drying stresses acquired during paper manufacture [5–7]. The much lower ZD values [8] reflect the compressibility and bonding level of the fibre layers in the paper structure.

The speed of sound in paper is most commonly measured using pairs of bimorph transducers placed lightly onto the surface of the test sheet. These are comprised of a metal paddle, a few millimetres wide, which is made to vibrate when an alternating voltage, of typical frequency 80 kHz, is applied to the piezoelectric crystals adhered on either side of the paddle. The paddles vibrate perpendicularly to their width producing a longitudinal sound wave that propagates along the surface of the sheet. Another transducer is used to detect the transmitted sound wave. Sound speed is calculated from the known distance of the spacing between the transducers divided by the time of travel of the sound wave pulse determined electronically. Commercial instruments, the Lorentzen & Wettre (L&W) tensile stiffness orientation (TSO) [9], Nomura Shoji SST 250 [10] or the SoniSys usually report the in-plane velocity squared, which is effectively 'E/ρ', known as the specific modulus or tensile stiffness index (TSI).

A popular application of sonic in-plane testing exploits the sensitivity of the method to the MD/CD ratio [11]. Paper machines are often several metres wide so that the web, in varying states of consolidation, is under tension. This, along with many open draws in the machine, all lead to a cross-machine reel profile in the MD/CD ratio in the resulting dry paper at the end of the reel. A highly varying MD/CD ratio of edge rolls, compared with rolls selected from nearer the centre of the machine, cause runnability issues once the paper is run through a printing or converting process [12], therefore it is desirable to minimise the MD/CD profile across a reel through a programme of iterative adjustments of headbox stock flows, stock jet to wire speed ratios, open draw tensions, drying strategies and so on.

An example is provided below where cross-reel strips from a fourdrinier paper machine producing linerboard were submitted for profile analysis. The cross-reel strips were supplied as rolls that were 3 ft or longer in the MD, crudely cut with a knife by walking along the length of the jumbo reel at the end of the paper machine. The strip is laid out flat on the floor, the edge of the roll is assumed to be the actual MD, perpendicular lines to the edges are drawn across the roll and test strips are cut across the reel at regular intervals. It is important to realise in profile studies that there is a variation in MD measurements as well as a CD. The objective is to observe whether a significant profile exists in the CD. In **Figure 3.1** below, a L&W TSO tester [9] was used to make four measurements in the MD at various successive positions across the reel. This instrument consists of a circular array of transducers that instantly provide measurements of in-plane V^2 upon electronic activation of alternating opposing pairs.

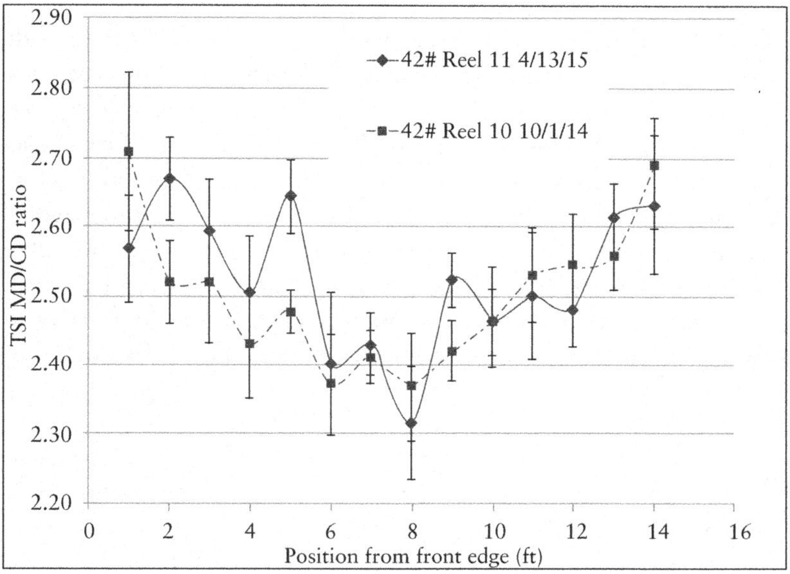

Figure 3.1 MD/CD stiffness profile for the 42# samples. The 'typical smile' MD/CD profile seen here is largely due to edge shrinkage and some edge flow. Error bars are standard deviations of 4+ measurements

The quantity of interest in reel strip profile measurements is the ratio of V_{MD}^2/V_{CD}^2, the MD/CD specific stiffness ratio. **Figure 3.1** shows the typical dip in the middle of the cross-reel strip that is seen in many fourdrinier-style paper machines that are not optimised to have a flat profile. The lower ratio in the middle, around position 8, is caused by the combination of stock edge flows on the forming fabric and drying shrinkage stresses [11]. The paper web in most paper machines endures many open draws while under MD tension and so inevitably shrinks in the CD during the manufacturing process. The contribution of edge flows may be assessed by examining the stiffness orientation angle [13], which is the clockwise measured angle of the maximum V^2 with respect to the assumed MD from the sample strip edge.

Orientation angle profiles in **Figure 3.2** show a slight rise or dip in the middle of the machine (positions 6 through 8) indicating an opportunity to flatten the profile by adjusting the stock to wire speed ratio on the paper machine, the so-called rush/drag ratio. Orientation is measured clockwise with respect to the assumed direction of the true MD, which is 0°. Negative values indicate or orientate towards the paper machine operator or control side, whereas positive values indicate or orientate towards the drive side. An overall negative profile suggests that there is too much recirculation flow causing a skewed overall flow towards the operator side of the paper machine.

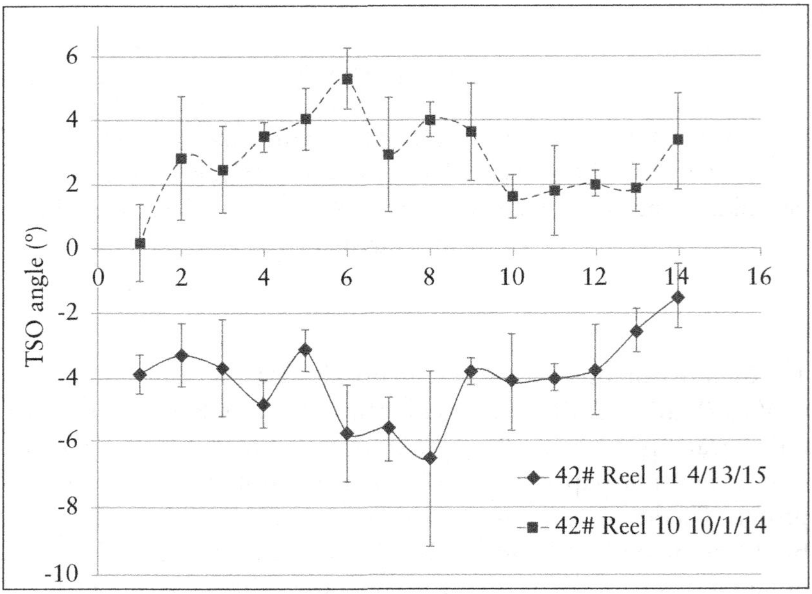

Figure 3.2 Corresponding TSO angles for the reel strips shown in **Figure 3.1**

3.2 ZD Ultrasonic Measurement

Speed of sound through the test sheet is used to determine the ZD modulus [14]. In this case, the transducers are 1 MHz driven piezocrystals coupled to plastic contacting delay blocks ending in neoprene sheets that contact either side of the sheet under a pressure of 50 kPa, the standard measurement, or 20 kPa for towel or tissue sheet measurements. A close-up of the transducers of one commercial development, the SoniSys instrument, is shown in **Figure 3.3**. This shows the central top and bottom ZD transducers which have black neoprene rubber contacting tips with concentric pairs of in-plane transducers behind. These transducers are made to rotate around the axis of the ZD transducer axis and drop down, *via* actuators, to take in-plane measurements at various angles on the test sheet. The action of the transducers is shown in **Figure 3.4**. When the orientation of the paddles are edge to edge, the propagation of the waves that are detected are shear waves. This has the rather esoteric application of calculating the in-plane Poisson ratios when the measurement is made at 45° with respect to the MD of the sheet [4].

Figure 3.3 Close-up photograph of the transducer contacting ZD and in-plane probes of the SoniSys instrument in their retracted vertical position. The test sheet is placed onto the back rubber mat

a) b) c)

Figure 3.4 Transducers in action in the SoniSys instrument. a) The ZD transducer is in contact with the sheet, b) the ZD transducer is retracted and the in-plane transducers are placed close to one another making a measurement and c) the in-plane transducers are taking a corresponding measurement spaced further part

The distance that is used to calculate the speed of sound through the sheet (ZD) is the sheet thickness or caliper. Paper caliper is affected by the combination of the sheet compressibility and surface roughness. For embossed towel and tissue, experimental evidence shows that caliper does not decrease appreciably once the measuring platen caliper is at 20 kPa or lower [15]. For dense printing, writing papers, the higher pressure of 50 kPa provides a measurement that correlates closely with the thickness of a stack of sheets [16]. Since paper is non-uniform on the scale of the longest fibres, a circular platen diameter is chosen to cover several flocs or visible clumps of paper, 20 to 30 mm is typical. Surface roughness is compensated by using soft rubber covers on the platens [17]. The rubber conforms to the surface undulations and provides a more accurate measurement of the caliper. This soft-platen measurement of caliper is known to provide what is termed the mechanical equivalent, or effective caliper, as shown by Setterholm [18].

The significance is that the soft-platen caliper corresponds to the same caliper ‘t’ that can also be calculated from first principles of mechanics, namely that the elastic modulus ‘E’ is related to the tensile stiffness ‘S_b’ as:

$$S_b = Et \quad (3.2)$$

Bending stiffness 'D' is the resistance of a sheet to bending. In the case of two-point beam bending, a specimen of a prescribed width and length is secured at one end and the force required to deflect the sheet test piece to a prescribed angle is measured using bespoke commercial testing instruments. The most common form of the instrument is the Taber version which measures bending moment [19]. Linear elasticity theory derives D in this circumstance to be:

$$D = \frac{Et^3}{12} \quad (3.3)$$

Combining the equation for D and S_b, the caliper that works is then:

$$t = \sqrt{\frac{12D}{S_b}} \quad (3.4)$$

Since the objective of ultrasonic measurements is to have measurements that correlate with physical properties, the 't' used in the calculations to convert sound speed to modulus is the soft-platen caliper [20], which is usually less than the caliper measured by hard platens by a few per cent.

3.3 Correlation of Ultrasonic Results with Mechanical Properties

Ultrasonic measurements conveniently correlate to many measured mechanical properties and in principle can be used as a quality check. The fundamental relationship that is used in this regard is between the S_b and the specific stiffness, or sound speed squared, and sheet basis weight 'β':

$$S_b = Et = \rho V^2 t = \frac{\beta}{t} V^2 t = \beta V^2 \qquad (3.5)$$

Many strength properties are related to the S_b. For example, as shown by Seth and Page [21] for well consolidated papers, the stress–strain curve shape is largely dependent on the modulus, as determined by the fibre quality, and only the strength values are affected by approximately 25%, as determined by the degree of fibre bonding. Therefore, sonic propagation can be used to assess sheet quality for a given sample set most easily by multiplying the basis weight of a sheet with the ultrasonically measured specific stiffness. In **Figure 3.5** below, a variety of paper and plastic film samples were tested for S_b using a universal testing machine and also for in-plane ultrasonic specific stiffness using a L&W TSO unit. There is a convincing correlation between the sonic calculated stiffness βV^2 and mechanical S_b. Typically, ultrasonic measured equivalents of physical constants are 30–50% higher than mechanically measured counterparts because of the viscoelastic nature of paper.

Many strength properties are dependent on 'E' and correspondingly the S_b, so that ultrasonics may be used instead of mechanical testing for a given sample set. In one example, unbleached softwood kraft pulp handsheets were prepared from slurries having undergone various dry strength additive treatments, the objective being to evaluate their effects in terms of standard short-span compression test or strength (SCT) [23] and out-of-plane crush [corrugated medium test (CMT)] [24]. The latter is a common qualifying test for corrugated board medium grades measured on test strips which are first sent through a laboratory corrugator to become fluted, the fluted strip is affixed to firm backing tape and then crushed out-of-plane on a compression tester. CMT is considered to be relevant to the flat crush resistance of corrugated board in shipping containers and SCT is the major contributor to the compression strength of corrugated board and boxes. The results are shown in **Figure 3.6**, where a good correlation of the physical results with mechanical values for compression strength is apparent.

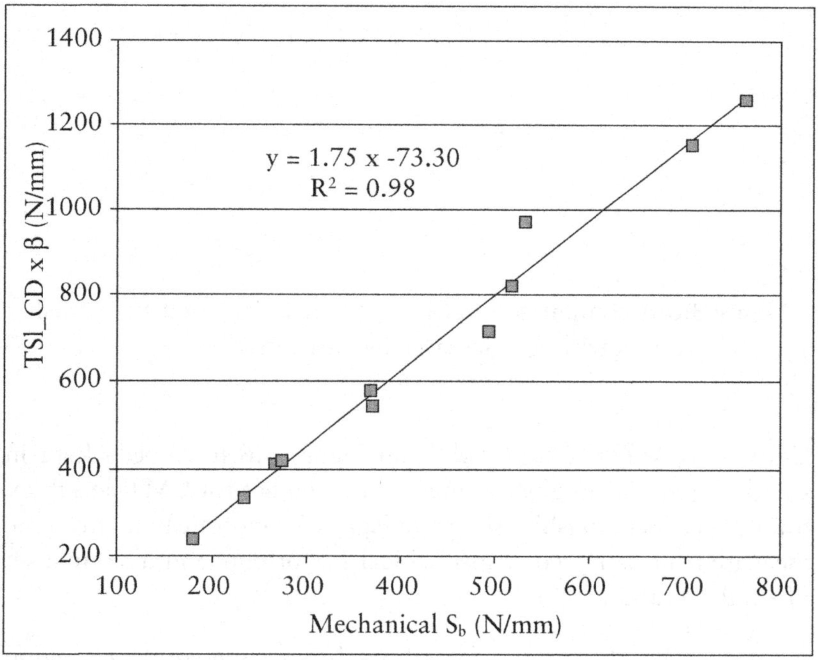

Figure 3.5 Results from a lab study using a L&W TSO instrument on a variety of paper samples and comparing the results to standard mechanical S_b. Reproduced with permission from R.E. Popil in the *Proceedings of TAPPI PaperCon 2010*, 2–5th May, Atlanta, GA, USA, 2010. ©2010, TAPPI [22]

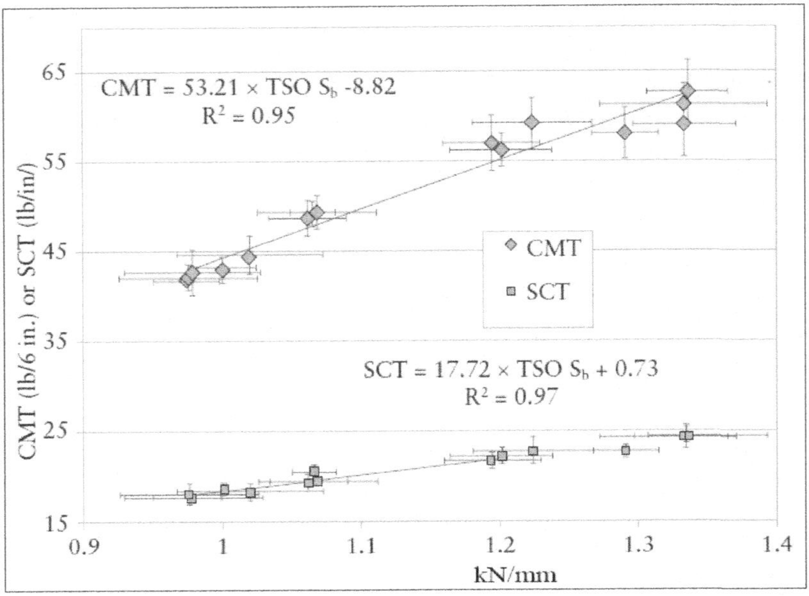

Figure 3.6 A comparison of out-of-plane corrugated strip crush results with S_b calculated as basis weight × V_{CD}^2

Units for CMT and SCT values in **Figure 3.6** have been kept in standard popular English formats of lb/6 inches for CMT and lb/in. for SCT. The relationships shown in **Figure 3.6** indicate that ultrasonic measurement can predict mechanical performance in a sample set of similar grades.

A range of linerboard and furnish mediums from southeast region USA mills were analysed for SCT and ring crush test (RCT) [25] compression tests. RCT is similar in concept to SCT, but also contains a component of bending failure [22], as it involves the compression of a ring structure as the test piece. Again, as shown in **Figure 3.7**, the mechanical strength results plotted against the CD TSO S_b obtained from the ultrasonics measurements, demonstrate good correlation. Note that the linear relationship between SCT and ultrasonic stiffness for the machine-made liner and medium set represented in **Figure 3.7** is different to the handsheet sample set shown in **Figure 3.6.** Note also that the RCT is lower in value than SCT although both are compressive strength. RCT being lower than SCT is attributable to RCT having bending of the test piece structure contributing to its failure. The advantage in using ultrasonics to test sheet quality is that no sample cutting or preparation is required as long as the sample can be placed into the testing instrument. The variation in ultrasonics measurements is often smaller, approximately 3%, compared with 7 to 10% variation for many mechanical measurements. The disadvantage is that the correlation between elastic sonic measurements and inelastic mechanical strength measurements varies if the composition of the sheets changes markedly *via* substantial changes in furnish or additives. Therefore, if used for routine quality screening, the correlation between ultrasonic stiffness and mechanical measurements should be checked whenever changes are made to the product under evaluatation.

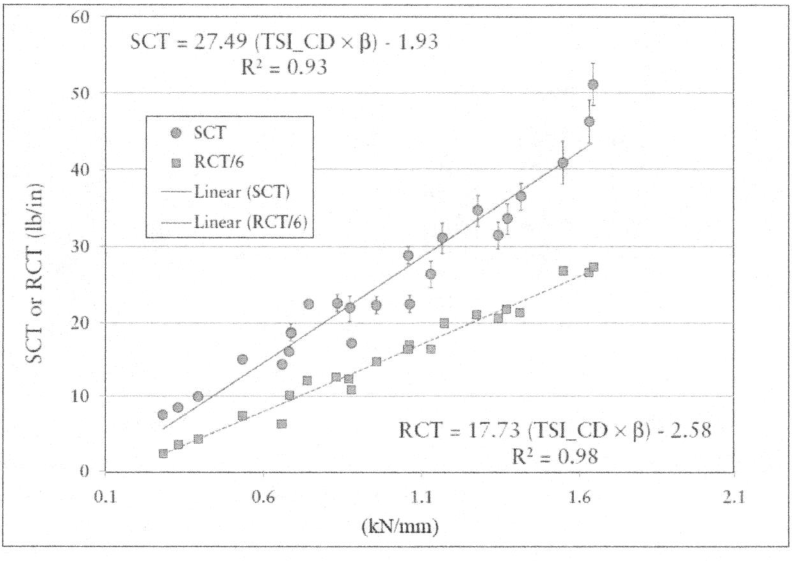

Figure 3.7 SCT and RCT for a selection of linerboard and corrugating medium of varying basis weights *versus* ultrasonic S_b

3.4 Application of ZD Ultrasonics

In the out-of-plane ZD, sonic propagation is affected by the degree of bonding between fibres [26]. Debonder additives are applied to towel or tissue products in an attempt to improve bulk, increase absorbency and apparent softness [27]. ZD sonic measurements are useful in assessing the potential efficacy of debonder treatments. In **Figure 3.8** below, a series of standard laboratory handsheets were prepared using various levels of a conventional debonder 'A' and two experimental debonders 'B' and 'C' at conventional application levels of 1.5, 3 and 5 lb/tonnes dry weight of fibre mixed into the stock slurry of a British Standard Sheet Machine handsheet former.

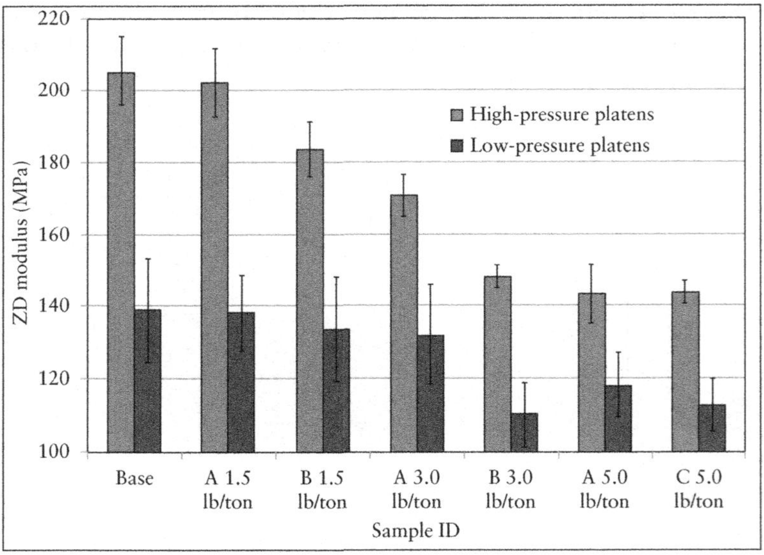

Figure 3.8 Ultrasonic ZD measurements at two different platen pressures for a series of handsheets prepared with differing levels of various debonders

The vertical axis in **Figure 3.8** shows the ZD modulus obtained using SoniSys equipment, where the time taken for a 1 MHz longitudinal wave to traverse the soft-platen caliper along with the measured basis weight produces the ZD modulus. There are two sets of ZD modulus data shown in **Figure 3.8**. One set is taken using a caliper pressure of 50 kPa and the other at a pressure of 20 kPa. The lower pressure produces less compression of the sheet so that overall, lower ZD moduli are obtained compared with the corresponding values obtained at the higher pressure. Both sets of data indicate that relative to the base sheet, which has no debonding additive, agent 'B' applied at the 3 lb/tonnes level can be expected to be the most effective treatment. Indeed, the decrease in ZD modulus was found to be accompanied by a significant measured increase in surface roughness, shown in **Figure 3.9**, and also a corresponding significant increase in compressibility, as measured by the difference in caliper at 50 kPa to that at 20 kPa. The ZD attenuation (loss of signal strength in dB from transmission) in the case of 20 kPa caliper pressure showed a good correlation with the measured Sheffield roughness.

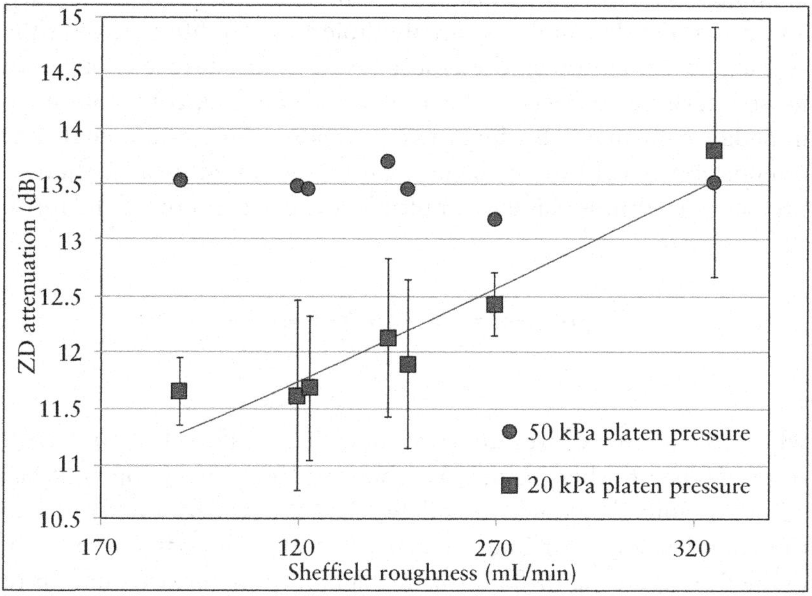

Figure 3.9 ZD attenuation *versus* Sheffield roughness for TAPPI handsheets with varying debonder dosage

Note that the 50 kPa ZD attenuation results show no correspondence with the increased surface roughness, suggesting that the 50 kPa caliper pressure suppresses the surface roughness compared with 20 kPa.

For this sample set, i.e., exploring the potential effects of debonding agents, no other significant differences were measured in strength properties, indicating that the ultrasonic measurements were actually more sensitive in detecting the expected effects of the chemical additives.

Since ZD sonic propagation is dependent on the coupling between fibres it can, in principle, detect tissue softness which is defined as the perception of softness by tactile feel using the thumb and fingertips of the hand. This is a contentious topic in itself, addressed by several researchers in the past and summarised by Hollmark [28]. The consensus is that softness comprises a combination of physical attributes such as low D, low surface friction, high compressibility, and a microscale roughness attributable to loose fibre ends on the surface. In sonic propagation, loose fibres on the surface, along with loose interfibre contacts in the absence of bonding, compromise an impedance mismatch leading to attenuation of the sonic signal. The attenuation A (dB) of the signal can be measured as a method to predict the softness using the multiple linear regression of the form:

$$\text{Softness} = aZ + bA/\beta + c\beta \quad (3.6)$$

where 'a', 'b' and 'c' are empirical coefficients 247.6, 19.1 and 0.82, respectively, or others as may be obtained from regression analysis, 'Z' is the impedance which is defined as the density multiplied by the sound velocity or equivalently, Z is also the basis weight 'β' divided by the time of flight of the signal, 'A' is the attenuation of the transmitted signal which is determined by Fourier analysis of the transmitted and received signals. An example of this application is shown in **Figure 3.10** where a selection of commercial paper towels were evaluated.

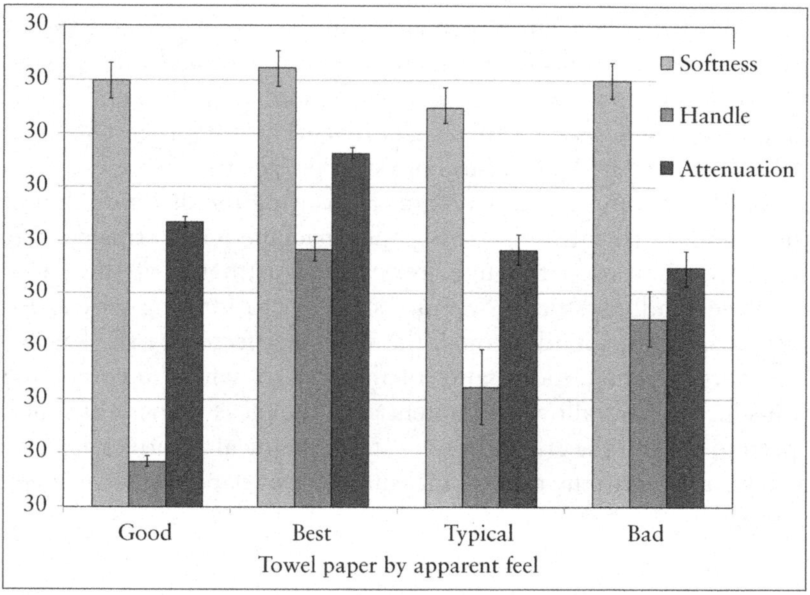

Figure 3.10 Comparison of the softness formula values, handle measurement and the attenuation for a towel sample set

Softness is calculated from the measurements using the regression formula and comparing the results with a tissue handfeel measurement, the so-called 'Handle-o-meter', and ZD attenuation. 'Handle' is measured as the force in grams required to insert the tissue with a sheet metal edge into a 20 mm or smaller slot and is also attributed as 'Drape' [29]. A lower value of handle is expected to reflect a softer feeling tissue product. For this sample set of paper towels, the distinction of softness, as calculated by the regression formula, is not significant, the handle better reflects the differences in basis weights of the sample set, however, the attenuation of the papers correlate well with the apparent feel of the sheets.

3.5 Summary

The speed of sound in paper is readily measured electronically using contacting transducers in commercially available equipment. The relationship between the velocity of sound squared and elastic modulus provides a convenient quality check of paper test samples that is often related to end-use physical properties of interest, such as tensile or compression strength. Measuring the directionality of the sound speed squared across paper machine-wide strips permits profile optimisation through iterative adjustments of the paper machine headbox stock flows and stock jet to forming wire speed ratios. Speed of sound through the sheet is affected by the level of intrafibre bonding and quality of the contact with the contacting transducers. This allows the potential of the measurement of sound speed to discern the effects of pulp stock chemical additives and can also be applied to the relative measurement of paper tissue or towel softness.

References

1. J.K. Craver and D.L. Taylor, *TAPPI Journal,* 1965, **48**, 3, 142.

2. J.F. Waterhouse, *TAPPI Journal*, 1994, 77, 1, 120.

3. R.W. Perkins in *Handbook of Physical Testing of Paper*, Eds., R.E. Mark and J. Borch, CRC Press, Boca Raton, Fl, USA, 2001, **1**, 2.

4. R.W. Mann, G.A. Baum and C.C. Habeger, *TAPPI Journal*, 1980, **63**, 2, 163.

5. T.R. Hess and P.H. Brodeur, *Journal of Pulp and Paper Science*, 1996, **22**, 5, 160.

6. D.W. Vahey, J.M. Considine, A. Kahra and M. Scotch in *Proceedings of Progress in Paper Physics Seminar*, 2–5th June, Otaniemi, Finland, 2008, p.2.

7. M. Titus, *TAPPI Journal*, 1994, **77**, 1, 127.

8. C.C. Habeger and W.A. Wink, *Journal of Applied Polymer Science*, 1986, **32**, 4, 4503.

9. G. Lindblad and T. Furst in *The Ultrasonic Measuring Technology of Paper and Paperboard*, AB Lorentzen and Wettre, Kista, Sweden, 2001.

10. T. Nomura, *Japan TAPPI Journal*, 1994, **48**, 1, 215.

11. S. Loewen and W. Foulger, *Pulp and Paper Canada*, 2002, **103**, 5, 42.

12. J. Paukku, M. Parola and S. Vuorinen in *Proceedings of the International Printing and Graphic Arts Conference*, 4–6th October, Vancover, Canada, 2004, p.245.

13. M. Santos, J. Perdigo and J. Velho, *Insight – Non Destructive Testing and Condition Monitoring*, 2007, **49**, 3, 146

14. C.C. Habeger, W.A. Wink and M.L. Van Zummeren, *Institute of Paper Chemistry Technical Paper Series*, 1988, August, No.301.

15. D.S. Keller, C. Feng and Y. Huang in *Proceedings of the International Paper Physics Conference*, 10–14th June, Stockholm, Sweden, 2012, p.32.

16. TAPPI T 411: Thickness (caliper) of paper, paperboard, and combined board (Revision), 1997.

17. W.A. Wink and G.A. Baum, *Institute of Paper Chemistry Technical Series*, 1983, March, No.133.

18. V.A. Setterholm, *TAPPI Journal*, 1974, **57**, 3, 164.

19. TAPPI T 489: Bending resistance (stiffness) of paper and paperboard (Taber-type tester in basic configuration) (Revision), 1992.

20. TAPPI T 511: Thickness of paper and paperboard (soft-platen method) (Revision), 2002.

21. R.S. Seth and D.H. Page in *The Role of Fundamental Research in Paper Making*, Mechanical Engineering Publication, London, UK, 1983, p.421.

22. R.E. Popil in *Proceedings of the Tappi Paper Conference*, 2–5th May, Atlanta, GA, USA, 2010.

23. TAPPI T 826: Short span compressive strength of containerboard, 2013.

24. TAPPI T 809: Flat crush of corrugating medium (CMT test), 2011.

25. TAPPI T 822: Ring crush of paperboard (rigid support method), 2011.

26. Y.C. Pan, C. Habeger and J. Biasca, *TAPPI Journal*, 1989, **72**, 11, 95.

27. J. Liu and J. Hsieh in *Proceedings of the TAPPI Papermakers Conference and Trade*, 16–19th April, Vancouver, Canada, 2000.

28. B.H. Hollmark in *Handbook of Physical and Mechanical testing of Paper and Paperboard*, Marcel Dekker, New York, NY, USA, 1984, **1**, 497.

29. O.C. Hansen, L. Marker, K.W. Ninnemannn and O.J. Sweeting, *Journal of Applied Polymer Science*, 1963, 7, 3, 817.

4 Bending Stiffness of Paper and Corrugated Board: The Connection to Caliper and Tensile Stiffness

Generally, the resistance of objects to a mechanical action such as applied stress is called stiffness. We have already encountered tensile stiffness (S_b) as the reaction or resistance of paper to tension. Similarly, the resistance of sheets or boards to bending action is termed bending stiffness(es) (D). D ties in the S_b and the caliper of the board or sheet and can, in principle, be calculated from tensile test measurements and caliper. D of paper is important to specify the paper rigidity required for printed tickets, business cards, folding cartons and in converting processes such as printing or folding. Conversely, low D is desired for towels, tissues and non-wovens where conformability or drape of the sheet products are desired features. D in corrugated boards limits the outward bulging of the side panels of boxes under load and contributes to the compression strength of boxes.

4.1 The Bending Elasticity Theory

The simple elasticity theory provides the formula for the D of a one-dimensional beam [1], which we can apply to the sheet of paper per unit width:

$$D = \frac{Et \times t^2}{12} \tag{4.1}$$

where Et, the elastic modulus 'E' multiplied by the effective caliper 't' [2], is the S_b obtained from the tensile test. The effective caliper removes the surface roughness contribution to roughness by using caliper-measuring platens in the standard gauge which are covered with soft neoprene. Measurement of paper D is achieved by applying a moment, the simplest example is the two-point application where a length of sheet is fixed at one end and a force is applied to the other end. The deflection that the test specimen undergoes in reaction to the force is related through the elasticity theory. For the two-point bending of a beam of length 'L' and width 'w', standard Euler mechanics elasticity analysis [1] provides the relationship:

$$D(mN-m)=\frac{60^{\circ}F(N)L^{2}(mm^{2})}{\pi w(mm)\alpha^{\circ}} \tag{4.2}$$

between the force 'F' subjecting a beam to a circular arc deflection of 'α' degrees. The commonly used Taber style of instruments measure the force required to achieve 15° of deflection of a 38 × 50 mm test specimen. This feature of Taber instruments is sometimes a source for confusion [3] as it reports the bending resistance in terms of the bending moment 'M', and not the D. The relationship between the moment and D is:

$$D=\frac{ML}{3\alpha w} \tag{4.3}$$

where 'L' is the span (50 mm), 'w' is the width (38 mm) and 'α' is the bending angle in radians (15°). It is useful to apply this formula to convert from Taber moment units 'M' in grams force-cm to D in mN-m, which is given below:

$$D(mN-m)=\frac{(M\times 9.807\times 10^{-2})(mN-m)L}{3\alpha w}$$

$$=\frac{M\times 9.807\times 10^{-2}(mN-m)\,50\,(mm)}{3\frac{\pi}{180^{\circ}}15^{\circ}\,38.1\,(mm)}$$

$$=0.1639\ \mathbf{M}\ (\text{grams force-cm}) \quad (4.4)$$

The formulas here all assume that beam strains are in the linear elastic region of the stress–strain curve for the material. This requires the assumption that the deflection is small, the moment point curves the beam around a circular arc and that the beam is long relative to its thickness, such that out-of-plane shear strain is negligible. Therefore, when testing samples, consideration must be given to limits in sample size and deflection to ensure linear elasticity.

The Taber [4] and Lorentzen and Wettre (L&W) bending resistance measuring instruments are shown in **Figure 4.1**. Both instruments typically use a 38 × 50 mm test specimen prepared using the punch cutter shown in **Figure 4.2**. The length of the test specimen is cut in the direction of interest to be measured, either in the machine direction of machine-made paper (MD) or the machine direction of paper (CD) for machine-made papers. The Taber turns the mounted sample, which is subjected to a counterweight, until a 15° deflection is attained. Different counterweight arrangements are applied according to a table that ascribes the counterweights required for the range of bending resistance being measured. Similarly, but simpler in design, the L&W instrument bends a clamped end of the test specimen through a rotation of 5° and measures the resulting force at the other end by a contacting load cell. Different specimen lengths and deflection angles can be selected for both instruments depending on the stiffness of the sample.

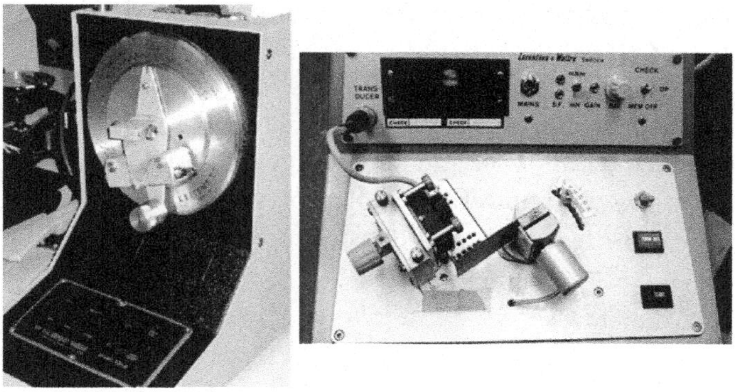

a) b)

Figure 4.1 The Taber a) and L&W b) two-point D testers commonly used in the paper industry to measure the bending of board and paper samples

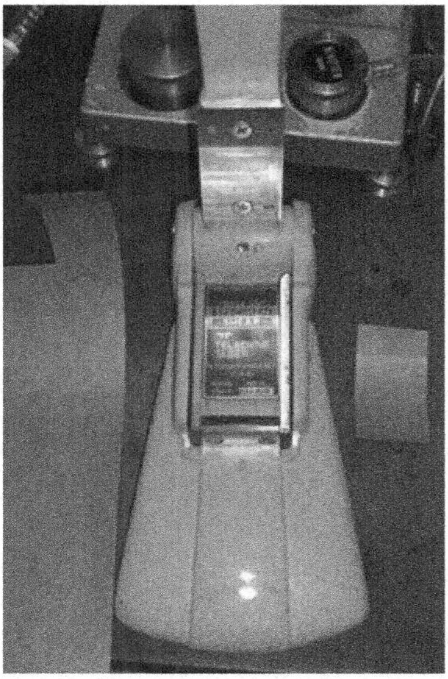

Figure 4.2 Both the Taber and the L&W bending measuring instruments use a punch cutter that prepares 50.2 × 38.1 mm test specimens, the length being along the MD or CD of the sheet

4.2 A Comparison of Different Methods of Measuring Bending Stiffness

An example of measuring the D of several samples and their relationship to S_b and soft and hard calipers follows. Seven different samples were tested and **Table 4.1** displays the designation, grade, basis weight, soft caliper and density for each sample used in this study.

Table 4.1 List of the various samples used in this S_b study, basis weights, soft caliper and density

Sample ID	Description	Basis weight (g/m^2)	Soft caliper (µm)	Density (kg/m^3)
A	42# Brown kraft linerboard	212	277	766
B	26# Neutral sulfite semichemical medium	130	191	679
C	Lightweight bleached kraft	75	92	823
D	Newsprint	45	57	808
E	Lightweight coated	47	44	1057
F	Mylar transparency film	146	98	1500
G	Synthetic paper	156	97	1610

Reproduced with permission from C.G. Carson and R.E. Popil, *TAPPI Journal*, 2008, 7, 12, 17. ©2008, Technical Association of the Pulp and Paper Industry [4]

These samples were used to acquire hard caliper, soft caliper, a 'stack' caliper, and the MD and CD calculated effective thicknesses. The hard and soft calipers were based on the Technical Association of the Pulp and Paper Industry (TAPPI) methods T 411 and T 551, respectively. The hard caliper was measured on an Emveco 200A and the soft caliper on an Emveco 210-DH caliper measuring instrument. The 'stack' caliper was measured by stacking 12 specimens of a sample and dividing the result by 12 (results of the measurements are given in **Figure 4.3**). The expectation of TAPPI T 411 is that the hard caliper result is a prediction of the thickness of a stack of specimens.

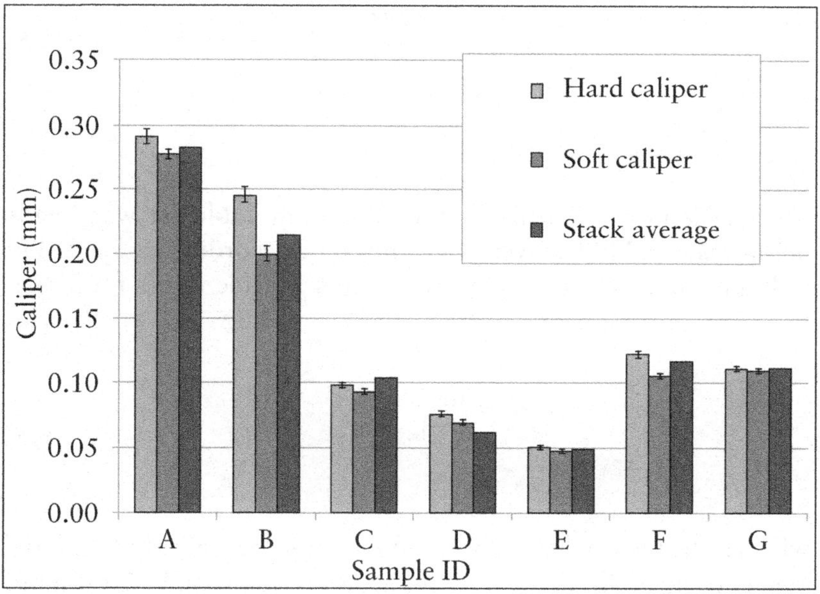

Figure 4.3 A comparison of hard and soft calipers for a range of samples listed in **Table 4.1**. Reproduced with permission from C.G. Carson and R.E. Popil, *TAPPI Journal*, 2008, 7, 12, 17. ©2008, Technical Association of the Pulp and Paper Industry [3]

The comparison of hard and soft calipers show the largest differences occur for samples of a low density, such as samples A and B (**Table 4.1**), which are also rougher surfaces compared with synthetic (sample G) or coated magazine journal paper (sample E).

The effective mechanical equivalent thickness 't_{eq}' was also calculated. This parameter, proposed by Setterholm, calculates the thickness from the relationship between S_t and S_b, which is derived as:

$$S_b = Et$$
$$D = \frac{Et}{12} t_{eq}^2$$
$$t_{eq} = \sqrt{12\frac{D}{S_b}} \tag{4.5}$$

where 'E' is the elastic modulus.

The tensile measurements for the samples in **Table 4.1** were made on an Instron 1122 universal test machine according to the TAPPI T 494 method. The tensile measurements provide values for S_b by:

$$S_b = \left.\frac{dF}{dx}\right|_{max} \times \frac{L}{w} \tag{4.6}$$

which is the Instron Series IX™ software S_t slope algorithm. 'L' is the gauge length (178 mm), 'w' is the width of the sample (25.4 mm), and 't_e' is the soft caliper measured and entered separately. D were measured on the L&W instrument using the default 5 deflection.

Figure 4.4 shows that the effective thickness, as calculated by the stiffness to D ratio, matches the measured caliper using the soft-platen method. Therefore, the soft-platen caliper can also be used to calculate the D once the S_b is known, as will be shown in the following. If we measure D using a different parameter or different instruments how can we know which is right? **Table 4.2** shows the D of the variety of samples described measured using three different methods.

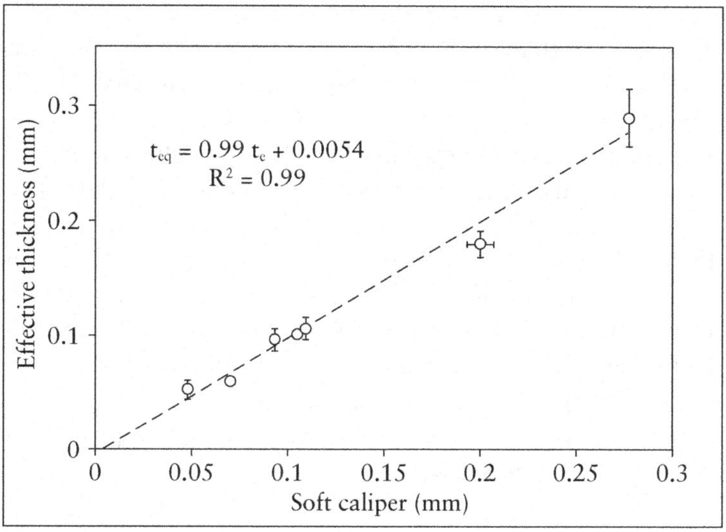

Figure 4.4 Comparison of the calculated effective Setterholm thickness with soft caliper measurements. Reproduced with permission from C.G. Carson and R.E. Popil, *TAPPI Journal*, 2008, 7, 12, 17. ©2008, Technical Association of the Pulp and Paper Industry [3]

Table 4.2 S_b results using different methods

Sample ID	MD			CD		
	5° L&W	30° L&W	Taber	5° L&W	30° L&W	Taber
A	9.235	8.106	8.932	5.051	4.466	4.800
B	2.278	1.966	2.012	0.891	0.812	0.799
C	0.358	0.352	0.333	0.144	0.141	0.144
D	0.102	0.099	0.101	0.016	0.017	0.017
E	0.065	0.057	0.058	0.031	0.030	0.027
F	0.431	0.469	0.465	0.417	0.437	0.448
G	0.393	0.409	0.398	0.360	0.390	0.384

Reproduced with permission from C.G. Carson and R.E. Popil, *TAPPI Journal*, 2008, 7, 12, 17. ©2008, Technical Association of the Pulp and Paper Industry [4]

Note, for example, that for the paper samples A–E, the stiffness at 30° is smaller than at 5°. The angles and spans for the samples were selected to obtain sufficient instrument sensitivity. The Taber instrument tests were all performed with its default 15° deflection. We will use calculations from the elasticity theory to check the accuracy of D measurements, which requires the assumption that the deflection is small, the moment point curves the beam around a circular arc and that the beam is long relative to its thickness, such that out-of-plane shear strain is negligible. These assumptions are not necessarily true during standard D measurements, so their conditions should be checked when invalidating a measurement. Stress/strain tensile test curves show that strains below 0.2% were within the linear elastic region of the curve. For the two-point method, the two-point bending strain 'ε' can be estimated as:

$$\varepsilon_e = \frac{\pi \times t_e}{120 \times l}\alpha^{\circ} \tag{4.7}$$

Table 4.3 shows a comparison of the angles, spans and equivalent strains used in the L&W measurements for the seven different samples. These particular parameters were selected to optimise accuracy. By comparison with stress/strain curves, we should expect the stiffnesses at 5° to be higher than those at 30°, since the elastic modulus at non-linear strains is lower than those at linear strains. Lightweight grades cannot be measured with the L&W instrument using default settings (50 mm span, 5° deflection), so large angle deflections are required at the risk of underestimating the S_b from non-linear strain.

Table 4.3 Listing of various test specimen S_b measurement parameters: free spans, resulting corresponding strains at two deflections and the Taber ranges used for the comparison of results

Sample ID	L&W spans (mm)		Bending strains ε (%)		Taber range
	MD	CD	5° deflection	30° deflection	
A	25	25	0.14	0.87	10–100
B	20	20	0.13	0.75	1–10
C	15	15	0.080	0.48	1–10
D	10	5	0.074/0.15[a]	0.44/0.89[a]	1–10
E	10	10	0.058	0.70	1–10
F	15	15	0.085	0.51	1–10
G	15	15	0.084	0.51	1–10

[a]The two numbers here correspond to the MD 10 and CD 5 mm spans, respectively.

Reproduced with permission from C.G. Carson and R.E. Popil, *TAPPI Journal*, 2008, 7, 12, 17. ©2008, Technical Association of the Pulp and Paper Industry [4]

Measurements of D listed in **Table 4.2** were compared with calculated D using the elastic modulus from tensile measurements by:

$$E = \left.\frac{dF}{dx}\right|_{max} \times \frac{L}{w \times t_e}$$

$$D = \frac{\langle E \rangle t_e^3}{12}\left[1 \pm \sqrt{\left[\frac{\sigma_E}{\langle E \rangle}\right]^2 + 3\left[\frac{\sigma_t}{\langle t_e \rangle}\right]^2}\right] \quad (4.8)$$

using the Instron Series IX™ software modulus algorithm with 't_e' the soft caliper measured and entered separately. Averages denoted by brackets were calculated based on five or more repeat measurements and the term in square brackets reflects the propagation of relative errors based on the standard deviations denoted by 'σ'.

Figure 4.5 compares the D of seven different papers measured using three different methods, i.e., L&W 5° bend angle, L&W 30° bend angle and Taber stiffness (15° bend angle), to the theoretical D calculated using the measured elastic modulus.

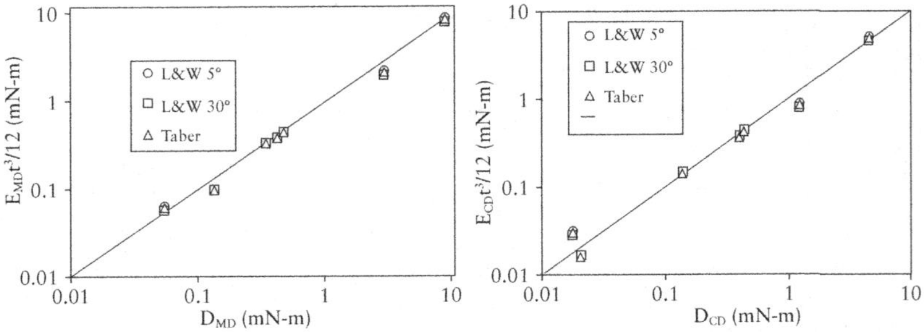

Figure 4.5 The theoretical D plotted against the experimental stiffnesses by three different methods: L&W 5°, L&W 30° and the Taber instrument. The results from all three methods correlate well to theory, including results of the lightweight grades. Reproduced with permission from C.G. Carson and R.E. Popil, *TAPPI Journal*, 2008, 7, 12, 17. ©2008, Technical Association of the Pulp and Paper Industry [3]

The results shown in **Figure 4.5** indicate that the various measurements are valid within experimental error when compared with the expected results from calculations. Thus, D measurements can be checked using the combination of tensile and soft caliper measurements utilising the equations from the linear elasticity theory.

4.3 Bending Stiffness of Corrugated Boards

D is important for corrugated boxes [5], as the amount of panel outward bulging that occurs during vertical load application needs to be limited. The term flexural rigidity [6] is often used for corrugated boards, although the relationship between flexural rigidity and D involves a Poisson ratio term which is usually not known nor measured for corrugated board. In the case of corrugated boards, the so-called sandwich beam theory for D applies to a good approximation:

$$D = \frac{E \times t \times h^2}{2} \tag{4.9}$$

where 'E' and 't' are the modulus and caliper of the outside linerboard and 'h' is the caliper of the board. In this case, hard caliper measurements suffice. The equation indicates that the medium does not contribute to the D [7] other than providing spacing between the linerboard facings producing the board caliper 'h'. More accurate calculations [8] show that the contribution of the fluted medium to D of the board structure is approximately 5%. The sandwich beam approximation is also useful for consideration of increasing the D of thicker paperboards, where greater gains are obtained when the outside facing layers are made to have high S_b.

The arrangement for testing the D of corrugated boards uses the four-point method shown schematically in the diagram of **Figure 4.6**. The load is applied at the ends of the test specimen by weights designated as '$\frac{P}{2}$'. The board is subjected to moments at either end from the weights '$\frac{P}{2}$' at a fixed distance, in this case 0.13 m, from the fulcrum points designated as triangles. The deflection of the board 'y' is related to the board's D through the relationship:

$$D = \frac{\left(\frac{P}{2}\right) 0.13\, L^2}{8wy} \tag{4.10}$$

'L' is the length of the board between the pivot points, typically 0.2 m cut along either the MD or the CD, 'w' is the width which is typically 0.1 m.

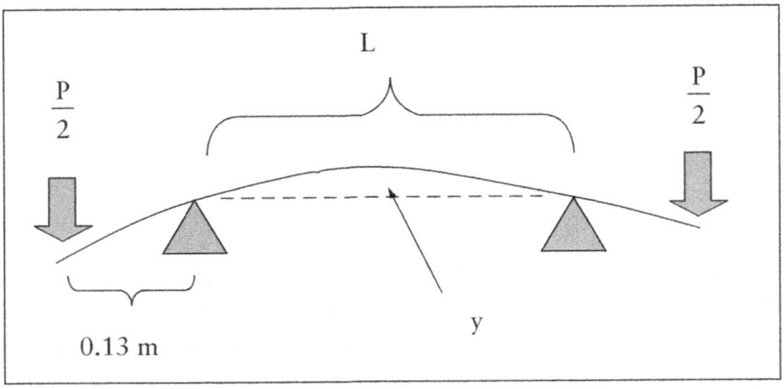

Figure 4.6 Schematic of the principle of the four-point bending arrangement for measuring D

The equipment is shown in **Figure 4.7**. The weights for '$\frac{P}{2}$' are supplied in incremental gram units.

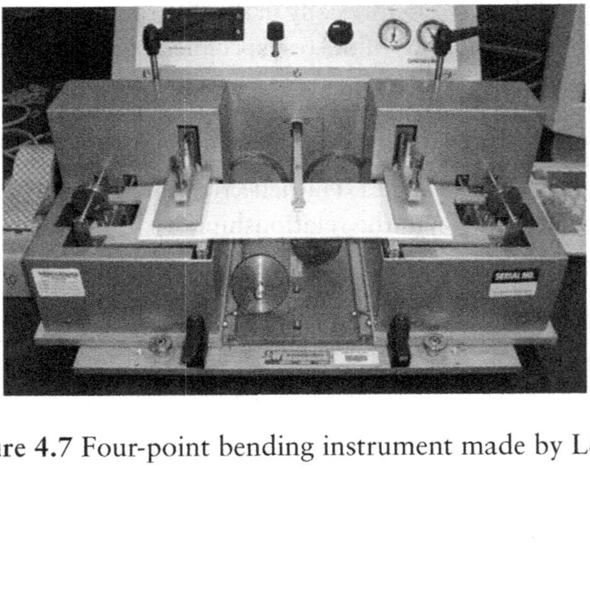

Figure 4.7 Four-point bending instrument made by L&W

So, for typical situations with boards cut into 12 × 4 inch pieces, the deflection 'y' panel read-out in mm, the convenient form of the formula becomes simply:

$$D(N-m) = \frac{0.0627\left(\frac{P}{2}\right)g}{y\ (mm)} \quad (4.11)$$

For C-flute corrugated board, '$\frac{P}{2}$' is 120 for CD and 220 for MD, and deflections are approximately 1 mm or less providing values of 15 N-m for the MD and 8 N-m for the CD.

The four-point bending method eliminates the effect of shear in the measurement [9] by the combination of the symmetry of load application and clamping restraint of the board test piece. In the alternative common three-point method [9], a test piece board is instead supported at either end and is simply deflected in the middle, and the load at the middlepoint is measured. For corrugated board, MD shear develops mostly from a relative displacement of the linerboard facings when the test pieces become bent. CD shear for corrugated board is considerbaly restricted by the facings that are adhered to the flute tips at the glue lines. Shear increases with shorter test pieces, and experiments show that test pieces have to be longer than 26" in the MD or longer than 10" in the CD to match the higher D values obtained with four-point bending.

4.4 A Comparison of Three- and Four-Point Bending Results for a Series of Boards

An example of the difference between four- and three-point bending results is shown in **Figure 4.8**. In this case, a heavyweight single-wall C-flute board was cut into lengths that were supported at two points 14" apart and deflected about 2 mm at the centre using the cross-head of a tensile testing machine. The cross-head was fitted with a load cell that measured the resulting force 'P' caused by deflecting the board. This provided the three-point D according to the equation (which neglects the shear contribution to bending):

$$D = \frac{PL^2}{48wy} \tag{4.12}$$

The same boards were tested using the four-point equipment. One series of boards was taken directly off the corrugator and cut to size on a computer-aided design (CAD) table, another series of boards was sent through the converting slitter-scorer, which makes the flaps of a box, and a third series consisted of samples cut from box blanks made by the folder-gluer.

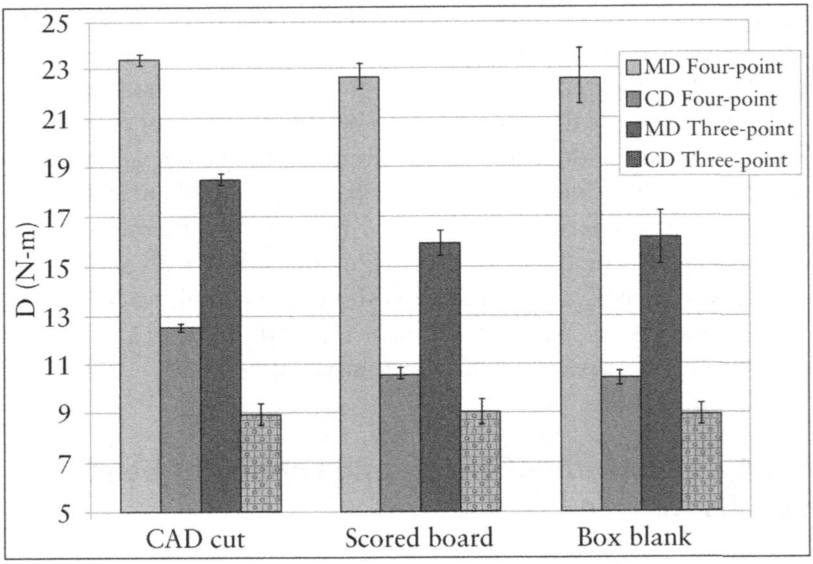

Figure 4.8 Comparison of four- and three-point D for a heavyweight C-flute corrugated board sample at three different points in the converting process as indicated. Error bars represent the 95% confidence interval of repeat measurements for each average value

All converting stages can introduce some degree of crushing of the board. In **Figure 4.8**, it is obvious that the D of the MD is greater than the CD by a factor of approximately two, reflecting the S_b orientation of the linerboard facings. The three-point D are lower than the corresponding four-point values by 30% or more. Boards that have been subjected to converting processes also show significantly lower MD values than the CAD cut board. This can be attributed to the effects of loss of shear stiffness caused by crushing of the board, which can also be detected by a correponding change in caliper measurement, as shown in **Figure 4.9**.

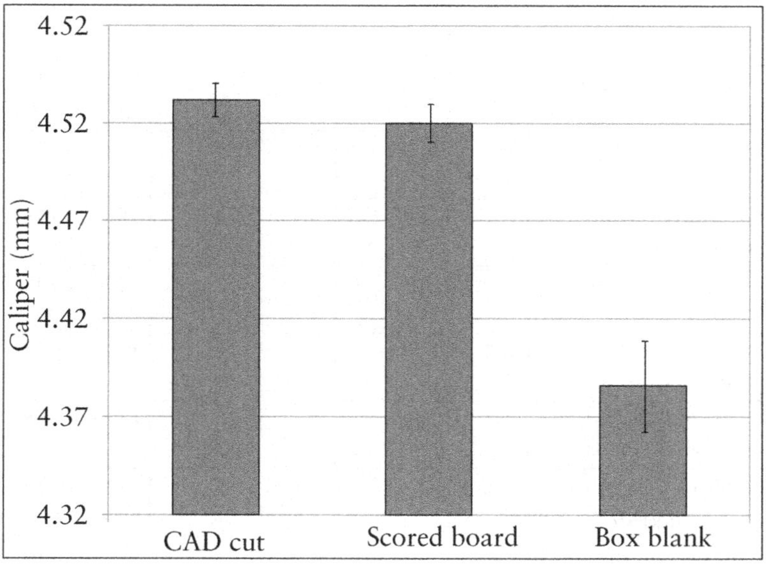

Figure 4.9 Comparison of the caliper of singe-wall board samples at different converting stages

Many manufacturers are interested in the possibility of replacing heavyweight single-wall board with lightweight double-wall board.

The idea here is that the loss in compression strength from reduction in basis weight can be compensated by an increase in D, as occurs with the increase in caliper of double-wall board. A description of the composition of the sample set are detailed in **Table 4.4** – boards A, H and I are single-wall, the rest are double-wall.

Table 4.4 Decription and composition of a test corrugated board sample set

Sample ID and description	Basis weight of board components (g/m^2)					
	Liner	Medium	Liner	Medium	Liner	Total weight
A – Heavyweight single-wall	337	112	337	–	–	786
B – Super lightweight double-wall	88	112	88	112	88	488
C – Lightweight double-wall	98	112	98	112	98	518
D – Lightweight B-B but with kraft	98	112	98	112	98	518
E – Medium weight double-wall	127	112	127	112	127	605
F – Heavyweight double-wall	151	112	127	112	151	653
G – Heavyweight B-B 'X' flute	151	112	49	112	151	575
H – Heavyweight single-wall kraft	274	112	274	–	–	660
I – Medium weight C-flute with kraft	254	112	254	–	–	620
J – Medium weight kraft B-B flute	161	112	161	112	161	707
B-B: Multi-walled board consisting of two B-flute medium layers						

The four- and three-point D were compared for this sample set. The four-point always measures higher than the three-point, particulary in the MD, as **Figure 4.10** shows. Some of the lighter linerboards used in preparing samples B and C, were made using novel headbox technology producing a sheet with no orientation, this results in the D of MD and CD not being significantly different from each other as is usually the case.

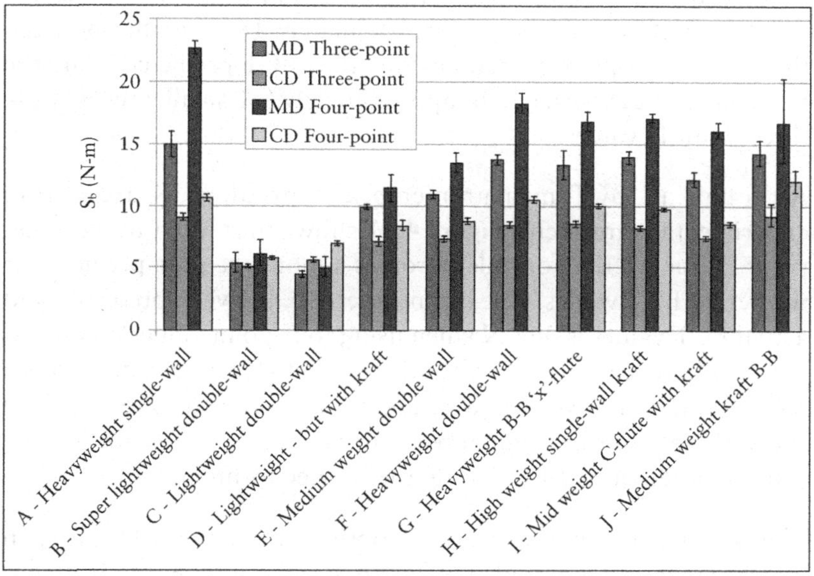

Figure 4.10 Comparison of three-and four-point D measurements for the sample set of corrugated boards

D of corrugated board and loss of shear stiffness from crushing are consequences of predicting the stacking strength [box compression test strength (BCT)] of vertically loaded boxes. The McKee equation for vertical BCT [10] is based on the analysis of the bending and failure of panels, and results in the semiemprical formula in metric form:

$$\text{BCT (kN)} = 0.375\ (\text{ECT})^{0.75}\left(\sqrt{D_{MD} D_{CD}}\right)^{0.25} Z^{0.5} \qquad (4.13)$$

with the board edge compression strength in kN/m, the MD and D of the board D_{MD} and D_{CD} in N-m, and the box perimeter (2 × {length+width}) Z in m. The geometric mean of the D term $\sqrt{D_{MD} D_{CD}}$ is raised to the ¼ power so that a 10% change in D will affect BCT only by approximately 2.5%. The change in the geometric mean D for the set of boxes averages approximately 18% smaller between three- and four-point measurements, thus BCT predicted using the formula can be expected to be approximately 5% smaller when using three-point D values.

Although all BCT measurements and predictions are within experimental agreement, **Figure 4.11** shows that when using three-point D, the McKee formula becomes a more accurate predictor of the actual BCT values. The mean difference between predicted and actual BCT values is 367 N when using four-point values compared with a difference of 181 N when three-point D is used instead. Since boxes subjected to vertical load are not constrained to restrict MD shear, three-point bending appears to be more relevant for a more acurate prediction of BCT using the McKee formula.

The question remains what is the optimal three-point D test span which is relevant as a predictor for a particular box. For long boxes, shear stiffness should not matter since three-point D is about the same as true four-point stiffness once the length exceeds 60 cm in the MD. In fact, a formula to convert from three-point 'D_3' measured at two different lengths 'L_1' and 'L_2' to four-point 'D_4', $L_1 > L_2$ is:

$$D_4 = \frac{D_{3,1} D_{3,2} (L_1^2 - L_2^2)}{D_{3,2} L_1^2 - D_{3,1} L_2^2} \qquad (4.14)$$

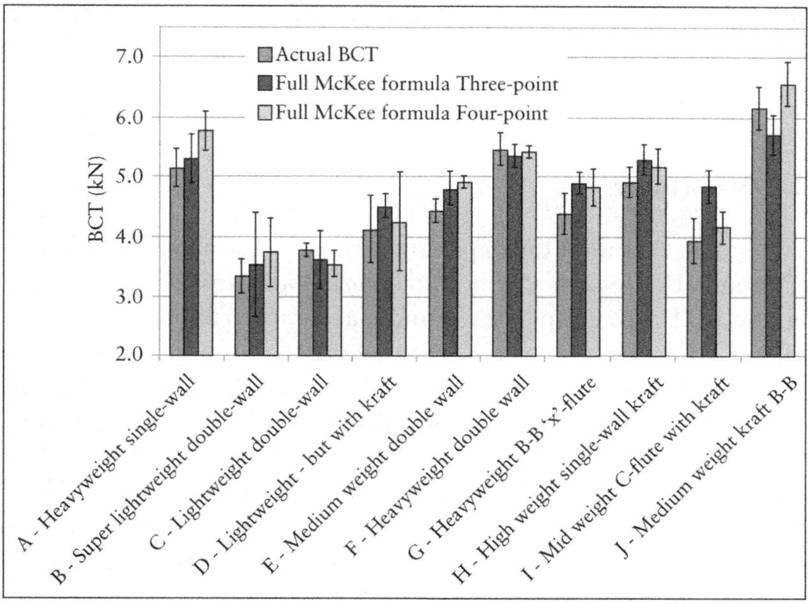

Figure 4.11 Comparison of actual BCT with the BCT predicted by the McKee formula using three- or four-point D values

For the case visited in **Figures 4.7** and **4.9–4.11**, the box dimensions were 61 × 41 × 66 cm, so that MD shear should not be significant, however, the selected test length of 36 cm still provides a better BCT prediction.

Deflection in D measurements should be limited to allow the assumption of linear elasticity theory to apply. In particular, the strain ε of the outerliner must be kept below 0.05%, which can be calculated using the formula:

$$\varepsilon(\%) = \frac{400 \times y\,(\text{mm}) \times t\,(\text{mm})}{L^2} \quad (4.15)$$

where the strain is given in percentage, the deflection 'y' in mm, 't' is the board caliper and 'L' is the test span between the two fulcrum points.

4.5 Summary

Two-point D is used for paper and some board materials, whereas four-point D is used for corrugated board. Both measurements depend on the S_b, 'E × t'. For paper, it is more accurate to use the soft-platen caliper if D as '$Et^3/12$' is to be calculated as a check of measurement accuracy. In this case, the modulus 'E' can be determined from the analysis of a tensile test and the calculation of '$D = Et^3/12$' can be compared with measurements of 'D'. For corrugated board, the D is approximated by the sandwich beam model '$Eth^2/2$', where 'Et' here is now the S_b of the outerliners and 'h' is the board caliper. Three-point bending provides smaller values than four-point, depending on the test piece length, because of the effect of shear. Three-point D values may provide a more accurate prediction of BCT when using the McKee equation.

References

1. S.P. Timoshenko and J.M. Gere in *Mechanics of Materials*, D. Van Nostrand, New York, NY, USA, 1972.

2. V.C. Setterholm in *Handbook of Physical and Mechanical Testing of Paper*, Volume 2, Eds., R.E. Mark and K. Murakmi, Macel Dekker, New York, NY, USA, 1984.

3. C.G. Carson and R.E. Popil, *TAPPI Journal*, 2008, 7, 12, 17.

4. TAPPI T 489: Bending resistance (stiffness) of paper and paperboard (Taber-type tester in basic configuration) (Revision), 1992.

5. T.M. Nordstrand, *Composite Structures*, 1995, 30, 4, 441.

6. TAPPI T 836: Bending stiffness, 4 point method, 2016.

7. R.E. Popil, *Bioresources*, 2012, 7, 2, 2553.

8. L. Carlsson, C. Fellers and P. Jonsson, *Das Papier*, 1985, **39**, 4, 149.

9. R.C. McKee, J.W. Gardner and J.R. Wachuta, *Paperboard Packaging*, 1962, **47**, 12, 111.

10. R.C. McKee, J.W. Gardner and J.R. Wachuta, *Paperboard Packaging*, 1963, **48**, 8, 149.

5 Compression Testing of Paper, Board and Boxes: Relationship to Tensile Testing, Elastic Modulus and the Influence of Artefacts

5.1 Introduction

Compression strength of paperboard is the dominating factor that governs the stacking strength of corrugated boxes. Considerable effort in the industry is focused on measuring the compression strength of boxes, corrugated board, and the linerboard and medium that comprise the corrugated boards. Compression strength is also considered in the design of folding carton packaging. Increasing the compression strength relative to the basis weight 'β' of the components remains a major development area. Success has been achieved through the wet pressing of sheets during the forming process to obtain high density and high strength. The use of strength additives in the stock also increases strength with the added expense of the chemicals and changes in dewatering properties of the stock. In addition, starch application at the dry end of the paper machine (PM) can provide some benefit, as can adjustment of the PM variables such as jet to wire speed ratio, open draw tensions, jet impingement angle and others.

Compression and tensile strengths (S_t) are related, although the mechanisms of failure are different. S_t failure is due to a combination of fibre fracture and fibre pull-out separation from the matrix. Close visual examination of a paper fracture edge will reveal long fibres extending out from the edge indicating fibre pull-out. Whereas compression failure consists of a combination of fibre buckling and separation of fibre layers at a visible crease, which is out-of-plane from the sheet. Both compression and S_t are related to the fibre modulus and the degree of fibre bonding; theory models exist [1] which can provide useful insight towards optimisation.

Shallhorn and Gurnagul [2] adapted the tensile Page equation [3] derived for S_t for compression strength modelling as a combination of fibre axial failure and fibre Euler buckling. They produced their analysis in the form of a harmonic average similar to the Page equation, analogous to resistors in parallel:

$$\frac{1}{\sigma_c} = \frac{1}{\sigma_o} + \left[\frac{Cw}{2\alpha E_f t^3}\right]\left(\frac{\rho_f}{\rho} - 1\right)^2 \qquad (5.1)$$

where the sheet compressive strength per unit basis weight at limiting high-density 'σ_o' (Nm/g) is related to the intrinsic fibre axial strength 'σ_f' (N) by:

$$\sigma_f = \frac{8}{3}C\sigma_o \qquad (5.2)$$

with '*C*' being the fibre coarseness (g/m), '*w*' the fibre width, '*t*' the fibre thickness, with t being much less than w, 'E_f' the fibre elastic modulus, 'α' is an efficiency factor ~1, 'ρ_f' and 'ρ' are the fibre (cellulose) and sheet densities, 'σ_c' is the sheet compression strength index (Nm/g) equated to the short-span compression test or strength (SCT), as measured by the Technical Association of the Pulp and Paper Industry (TAPPI) T 826 method [4]. Inspection of the equation shows that compression strength decreases with fibre wall thickness '*t*', increases with wet pressed sheet density 'ρ', and refining is more effective at increasing compression strength for a given increased density compared with wet pressing, as refining also decreases fibre coarseness. Significant differences in σ_o were noted, for example, between spruce fibres (E_f = 56 GPa) and coarser thicker Douglas fir fibres (E_f = 48 GPa). At low-density fibre buckling dominates, therefore a larger fibre modulus and larger fibre thickness increase compressive strength. At high density, fibre compressive strength dominates, hence a low fibril angle (high E_f) and low fibre thickness produce a larger compression strength. Fibre modulus is related to the sheet elastic modulus 'E_p' through the relationship derived by Page and Seth [1] for randomly oriented sheets:

$$E_P = \frac{1}{3} E_f [1 - k/RBA] \tag{5.3}$$

where 'k' is a constant that is specific to a pulp and RBA is the fibre-relative bonded area, which can be determined by optical scattering coefficient measurements. Experimentally, Shallhorn and Gurnagul inferred E_f as being just $E_p/3$ with sheet modulus E_p obtained at the highest level of wet pressing, i.e., 7 MPa, the assumption is that at this pressure the sheet elastic modulus is close to its maximum.

Figure 5.1 shows a hypothesised superposition of compression and tensile stress–strain data for a typical fourdrinier-made linerboard sample. At low strains, i.e., 0.5% and lower, the response of the sample to compression or tensile strain is linear and to a first approximation is described by Hooke's law:

$$\sigma = E\varepsilon \tag{5.4}$$

σ is the stress, i.e., the applied force divided by the width multiplied by the effective caliper of the test strip, ε strain is the displacement of the test strip divided by its initial free length and 'E' is Young's modulus in this approximation. For an orthotropic solid such as paper, the strains, stresses and modulus are all directional, here we are principally concerned with the cross machine direction of paper (CD) direction and the directional notation, in Hooke's law, is dropped. Handsheets are randomly oriented and so the MD CD notation does not apply in this case as well. The modulus at low strains is the same whether the sample is under compression or tension, as **Figure 5.1** indicates.

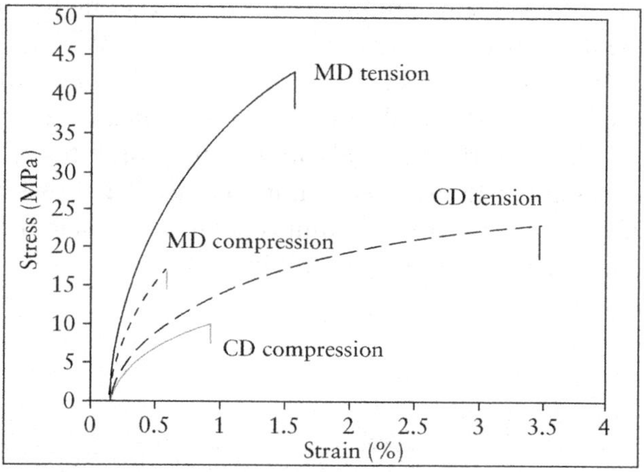

Figure 5.1 Superimposed stress–strain data for tensile and compression tests for a typical machine-made linerboard sample

Machine-made paper (MD)/CD modulus and strength ratios are approximately the same, about 2 for fourdrinier papers. S_t are also about twice as large as the corresponding compression strengths. At high levels of sheet consolidation and bonding, the S_t and compression strength properties are related and determined by the fibre quality and somewhat affected by the degree of bonding. The relationship between compression and tensile measurements allows the opportunity, in some circumstances, to estimate the compression strength of papers from tensile tests when compression tests are unavailable or impractical to perform. The compression testing of paper becomes subject to artefacts when dealing with lightweight papers, as buckling of the test specimen can produce an anomalously low compression strength value. In such cases, if tensile data are available for higher basis weights, the correspondence between S_t and compression strengths can be extrapolated to infer the compression strength at low basis weight.

An example of this follows, using the ring crush test (RCT) [5], SCT and tensile test results in combination. The marketing of linerboard and corrugating medium predominantly use the RCT as a quality criterion. This consists of cutting a strip ½ × 6 inches, the length being in the MD of a sheet, placing the test strip in a circular grooved fixture and crushing the assembly in a compression tester with the platen advancing at ½ inch per minute. **Figure 5.2** shows a sample in the RCT fixture where the compression tester platen has been allowed to progress beyond the peak load.

Figure 5.2 Photograph of an RCT sample of high basis weight after the test. Reproduced with permission from R.E. Popil in the *Proceedings of the TAPPI PaperCon 2010*, 2–3th May, Atlanta, GA, USA, 2010. ©2010, Technical Association of the Pulp and Paper Industry [6]

A distinct repetitive buckling pattern going along the length of the sample is evident in **Figure 5.2**, as is the top edge having rolled due to contacting the compressing platen. Although the diameter of the ring is some 50 mm and the height of the sample is 12.5 mm, it can be shown from simple mechanics analysis that such a short paper cylinder is still subject to buckling when placed under vertical load. The buckling load 'σ_{cr}' of a thin shell ring, based on the analyses of Timoshenko and Gere [7], is:

$$\sigma_{cr} = \frac{E_{CD}}{\sqrt{1 - \nu_{12}\nu_{21}}}\frac{t}{R} \tag{5.5}$$

where E_{CD} is the CD modulus, 't' is the thickness of the test specimen, 'R' is the radius of the ring fixed at 24.2 mm, the Poisson ratio term under the square-root sign adds a few per cent correction to the buckling load and does not vary significantly with different papers. The criterion for this equation to be valid is that the ratio of the buckling column height 'l' over R must be sufficient to fit at least one buckling wave and is stated as:

$$l = 1.72\sqrt{Rt} \tag{5.6}$$

This criterion is fulfilled for the standard sample of t = 0.3 mm and RCT strip width of 12.4 mm, of which 6 mm protrudes outside of the fixture base, so buckling in the RCT test can be expected from a theoretical perspective. Buckling of a sample is undesirable as it will lower the peak load when a sample is tested under compression. Accordingly, a different geometry for compression testing has been devised in the SCT.

The underlying principle in the SCT test method is that the sample is constrained by clamps eliminating any buckling during the test.

If the test sample strip under compression is considered as a simple elastic beam, then peak stress by buckling from load applied to the two ends of the beam is, from first principles, given by:

$$\sigma_{max} = \frac{k\pi^2 E_{CD}}{\lambda^2} \qquad (5.7)$$

where E_{CD} is the modulus of the test sheet in the CD, which is the usual test direction of interest, k is a constant depending on the end-point conditions and 'λ' is the important slenderness ratio defined as:

$$\lambda = 2\sqrt{3}\frac{l}{t} \qquad (5.8)$$

where 'l' is the length of the sample strip and 't' is its thickness. Plots of σ_{max} *versus* λ show that the compression stress stabilises once λ <30. For linerboards, this translates to strip lengths of approximately 1 mm and to accommodate the range of linerboards, the standard has been set with the free unclamped strip length for a compression test being 0.7 mm. Thus, the SCT test has 15 mm wide strips clamped at two ends with a free span of 0.7 mm, a fixed-load cell attached to one clamp measures the peak force once the opposite clamp is advanced. The point of this arrangement is to eliminate any buckling of the test specimen and thus obtain the compression strength without any sample bending. Unfortunately, testing such a small sample area of 15 × 0.7 mm subjects the values to a comparatively high variability related to the samples' mass non-uniformity on the mm scale, also commonly called paper 'formation'. In fact SCT, like many strength characteristics, is proportional to basis weight. Accordingly, a high variation in basis weight can be expected to also result in a high variation in strength values.

A comparison of SCT values to RCT is instructive to illustrate the influence of bending in the latter. A sample set consisting of 18 softwood linerboards and medium manufactured in the US southeast were tested for RCT, SCT, basis weight, caliper and ultrasonic stiffness using the Lorentzen & Wettre (L&W) tensile stiffness orientation (TSO) instrument [8]. The results are presented in **Table 5.1.** The buckling to compression stress ratio is calculated as:

$$\frac{\sigma_{cr}}{\frac{SCT}{t}} = \frac{E_{CD} \times t}{R} 1.05 \quad (5.9)$$

since SCT is given per unit width, dividing SCT by the caliper produces the compression stress, 'R' is the ring radius of 24.2 mm, the Poisson term of **Equation 5.5** '$\sqrt{(1 - \nu_{12}\nu_{21})}$' has been approximated to be constant at ~0.95 and '$E_{CD} \times t$' is the tensile stiffness (S_b) provided by the L&W TSO in **Equation 5.10** as:

$$E_{CD} \times t = (TSI_{CD} \times \beta)\ 0.571 + 41.8 \quad (5.10)$$

where 'β' is the basis weight and the factor 0.571 with offset 41.8 accounts for the difference between ultrasonic stiffness and mechanical S_t obtained by a separate experimental correlation [6].

A plot of the critical buckling to compression stress ratio is shown in **Figure 5.3,** where it is evident that buckling can be expected for basis weights less than 81 g/m^2, as the ratio is less than 1 at basis weights below 81. Shallhorn and Gurnagul [10] noted that the elasticity theory formula for the critical buckling stress for thin-walled tubes is an overestimation by a factor of approximately 2 compared with experimental data. This then brings the basis weight where the critical buckling stress becomes less than compression stress to approximately 180 g/m^2, this latter estimate is plotted as a dashed line in **Figure 5.3**.

Table 5.1 Data for a set of linerboards and medium used for commercial corrugated board in the southeastern US

Sample	Caliper (mm)	Basis weight (g/m^2)	TSI_CD (km/s)^2	CD SCT (N/mm)	RCT_{CD} (N/mm)
Liner A1	0.232	150.1	4.59	3.26	1.79
Liner A2	0.268	171.1	4.88	3.93	2.22
Liner A3	0.305	195.3	5.46	3.86	2.95
Liner A4	0.392	249.5	4.7	5.39	3.45
Liner A5	0.420	275.5	4.9	5.47	3.57
Liner A6	0.476	293.9	4.69	5.86	3.78
Liner B1	0.600	370.4	4.43	8.06	4.62
Liner B2	0.427	277.1	5.13	6.35	3.69
Liner B3	0.725	437.1	3.78	8.96	4.71
Liner B4	0.327	206.3	5.51	4.59	2.87
Liner B5	0.663	393.1	3.96	7.14	4.67
Liner B6	0.260	167.0	5.24	3.80	2.17
Liner B7	0.524	323.8	3.97	6.03	3.66
Liner B8	0.229	149.6	4.55	2.81	1.81
Liner B9	0.322	195.8	5.43	5.01	2.88
Liner B10	0.319	205.4	4.68	3.88	2.56
Liner B11	0.203	128.9	5.13	2.50	1.14
Liner B12	0.259	177.6	4.97	3.00	1.92
Medium 1	0.146	78.5	3.58	1.32	0.44
Medium 2	0.192	88.3	3.76	1.50	0.64
Medium 3	0.197	95.7	4.12	1.75	0.74
Medium 4	0.208	126.9	4.21	2.62	1.31
Medium 5	0.251	163.7	4.57	3.88	2.16

TSI: Tensile stiffness indices, CD and MD

Reproduced with permission from R.E. Popil, *BioResources Journal*, 2012, 7, 2, 2553. ©2012, North Carolina State University [13]

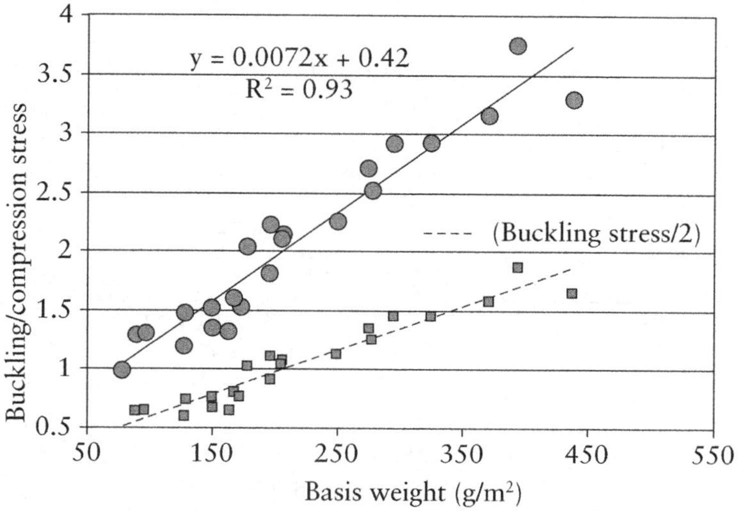

Figure 5.3 Ratio of the calculated buckling to compression stress for the sample set of commercial linerboards and medium listed in **Table 5.1**. The squares represent the ratio with the calculated buckling stress divided by 2. Reproduced with permission from R.E. Popil in the *Proceedings of the TAPPI PaperCon 2010*, 2–3th May, Atlanta, GA, USA, 2010. ©2010, Technical Association of the Pulp and Paper Industry [6]

Therefore, buckling of samples under compression can be expected for low basis weights, possibly as much as 180 g/m^2. Buckling of samples at low basis weight should produce a lower compression strength than that provided by SCT. This is shown in **Figure 5.4** for the sample set presented in **Table 5.1,** where the units for RCT have been converted to be the same as SCT, i.e., units for both are kN/m or equivalently, N/mm.

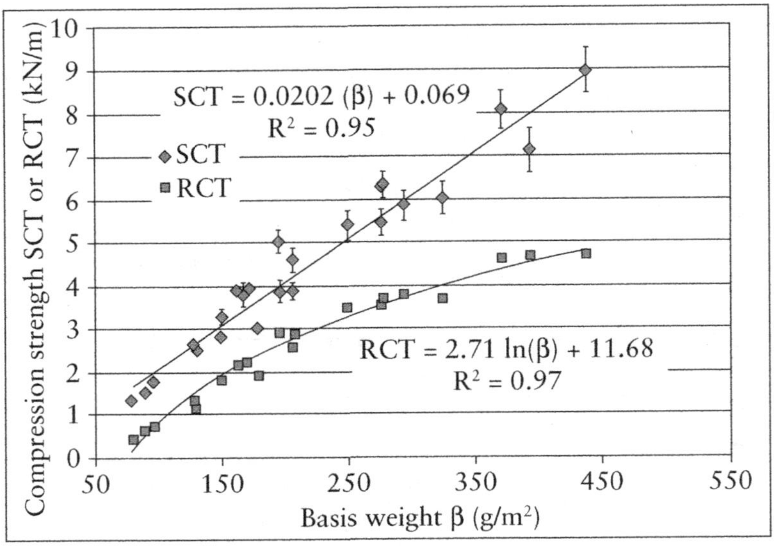

Figure 5.4 Comparison of SCT and RCT results for the sample set in **Table 5.1**. Reproduced with permission from R.E. Popil in *Proceedings of the TAPPI PaperCon 2010*, 2–3th May, Atlanta, GA, USA, 2010. ©2010, Technical Association of the Pulp and Paper Industry [6]

Figure 5.4 shows that at basis weights below 200 g/m^2, SCT is larger than RCT by approximately 1.7 times. The higher SCT values can be attributable to the role of bending, which lowers the RCT. At higher basis weights, the difference between SCT and RCT increases and the upper free edge of the test strip becomes rolled due to contact with the advancing platen of the compression tester. The best fit regression analysis in **Figure 5.4** shows that SCT is linear with basis weight, whereas the RCT relationship best fit is logarithmic with basis weight. Nonetheless, RCT is still a preferred specification for the marketing of corrugated board component linerboards or medium, as the variability of RCT is approximately two-fold lower than for SCT. The results in **Figure 5.4** show error bars representing 95% confidence intervals for 10 measurements per sample. The coefficient of variation, expressed as a percentage ratio of the standard deviation to the average value, averages 7.8% for SCT compared with 4.7% for RCT. The lower variability for RCT allows the production of linerboard at a lower basis weight that will meet a specified marketing specification. However, mill producers of linerboards prefer SCT as a quality criterion, as it appears to be more easily controlled through papermaking operations such as pulp stock refining, wet pressing density, paper basis weight, softwood/hardwood ratio, stock additives, and so on.

Besides being subject to buckling, rolling edges and not tracking basis weight, RCT is also two-sided, meaning results can be significantly different depending on the side facing the inside of the cylindrical ring undergoing vertical compression. The majority of corrugating medium papers are made on fourdrinier PM such that the drainage of the stock on the forming wire is one-sided, leaving the top side more filled with filler and fines than the wire side. The top side, when fines interfere with fiber bonding, can be somewhat weaker than the wire side. If the cylindrical buckling of the test piece is primarily outward, then the outside of the ring is subjected to tension and the inside to compression. **Figure 5.5** shows the results of 18 samples of medium from two paper machines at a mill where three measurements were made for each sample with the top side of the sample being inside the test ring and three outside of the ring.

In most of the cases shown in **Figure 5.5**, RCT is a few per cent higher and sometimes significantly so when the stronger wire side is on the inside of the buckling test cylinder. RCT is also sensitive to the cutting method. Parallel sides must be maintained so it is best to use a punch cutter rather than a guillotine-style cutter. Furthermore, test strips must be punched one at a time rather than several sheets at a time, otherwise edges are obtained that curl when tested, leading to lower than expected RCT values.

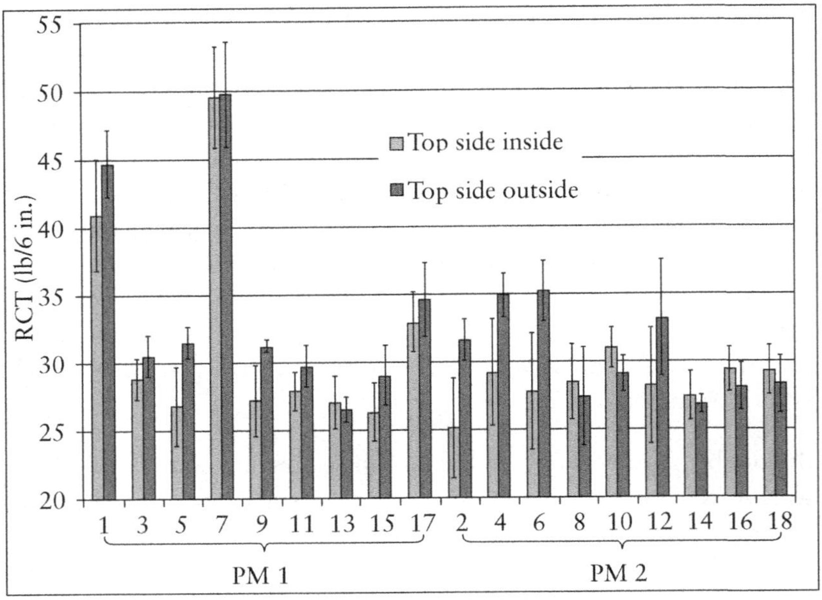

Figure 5.5 Comparison of RCT (in standard English units of lb/6 in.) for 18 samples with different side orientations as indicated (1 lb/6 in. = 0.0292 KN/m)

To satisfy long-established marketing requirements that specify RCT, with its combined bending and compression, and to connect it to the less artefact-prone SCT, a model can be devised based on the empirical result for structures that fail through combined compression and bending with the form for the maximum load 'P_{max}' as:

$$\frac{P_{max}}{P_{cr}} = \alpha\left(\frac{P_{com}}{P_{cr}}\right)^b \quad (5.11)$$

where 'P_{cr}' is the critical buckling load, 'α' and 'b' are constants, and 'P_{com}' is the intrinsic compression strength of the material. Thus, taking RCT for P_{max} and SCT for P_{com}, the proposed model for RCT becomes:

$$RCT = \alpha(SCT)^b(\sigma_{cr})^{1-b} \tag{5.12}$$

The availability of ultrasonic TSO instruments [8] on automated testing equipment installed in many mills makes use of the above equation in a convenient way to obtain an RCT value. Since the $(E \times t)$ term for the formula for the buckling stress 'σ_{cr}' is equivalent to $TSI_{CD} \times \beta$, the 'R' is constant in the RCT, we obtain by substitution into the previous formula:

$$RCT = C(SCT)^b(TSI_{CD}\beta)^{1-b} \tag{5.13}$$

with 'C' being a new constant. Keeping the units shown in **Table 5.1**, RCT and SCT in N/mm, TSI_{CD} in $(km/s)^2$ and β in g/m^2, a three parameter fit of the model equation to the data determined the model constants to be:

$$RCT = 0.0326\,(SCT)^{0.46}(TSI_{CD}\beta)^{0.54} \tag{5.14}$$

which produces an average residual error of 0.177 N/mm compared with a simpler linear model fit of RCT = 0.596 × (SCT), which produces an error of 0.246 N/mm, as shown in **Figure 5.6.**

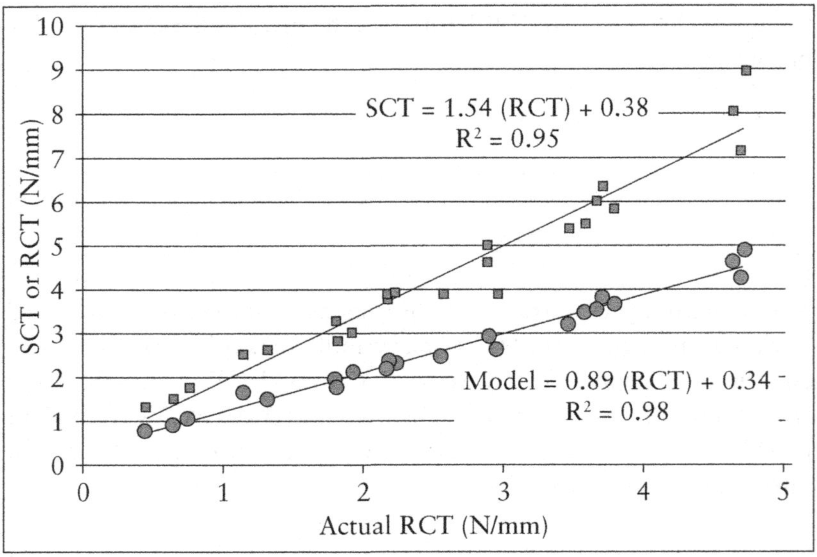

Figure 5.6 Plot of the RCT model fit (round points) and the SCT values (square points) *versus* actual RCT for the samples listed in **Table 5.1**

The same method was applied to another set of challenging lightweight linerboards consisting of 100% recycle furnish. The same analysis was applied but yielded different results, as shown in **Figure 5.7.**

The basis weight and SCT data in **Figure 5.7** were obtained from an automated testing machine. The RCT results in the plot came from manual laboratory testing but is preferably replaced by automated testing or otherwise calculated from available data using the existing automated instruments. Note that, as before, despite the fact that both measurements describe compression strength, RCT is always less than SCT, ranging from 3.5 times lower at a low basis weight to 1.3 times lower at a higher basis weight.

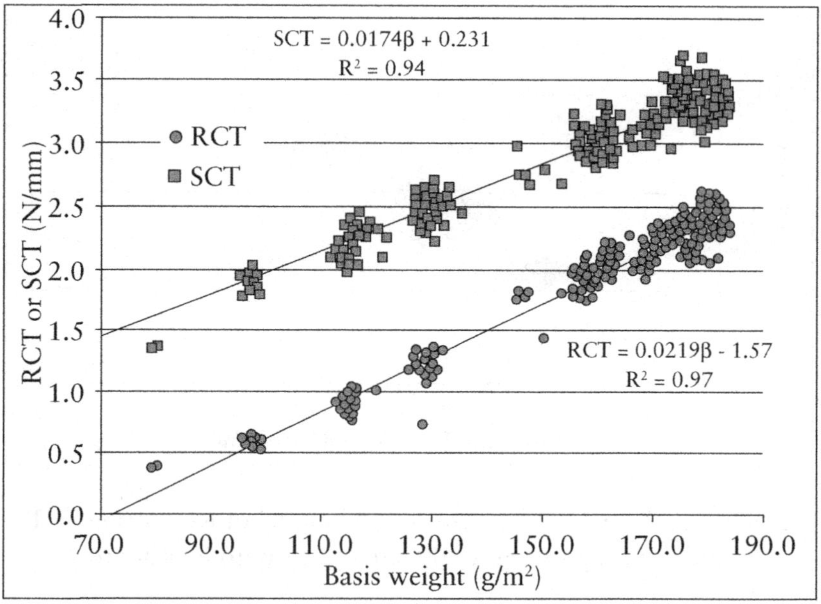

Figure 5.7 A plot of SCT and RCT *versus* basis weight for samples from a 100% recycle mill

Therefore, a similar analysis was undertaken as before to find a predictive relationship for SCT using a two parameter fit model. The least amount of scatter and best correlation for this particular data set was found to use a combination of SCT and S_t data. S_t in the CD, dubbed ($S_{t,CD}$), like S_b, can also be expected to be related to the CD elastic modulus for a given sample set.

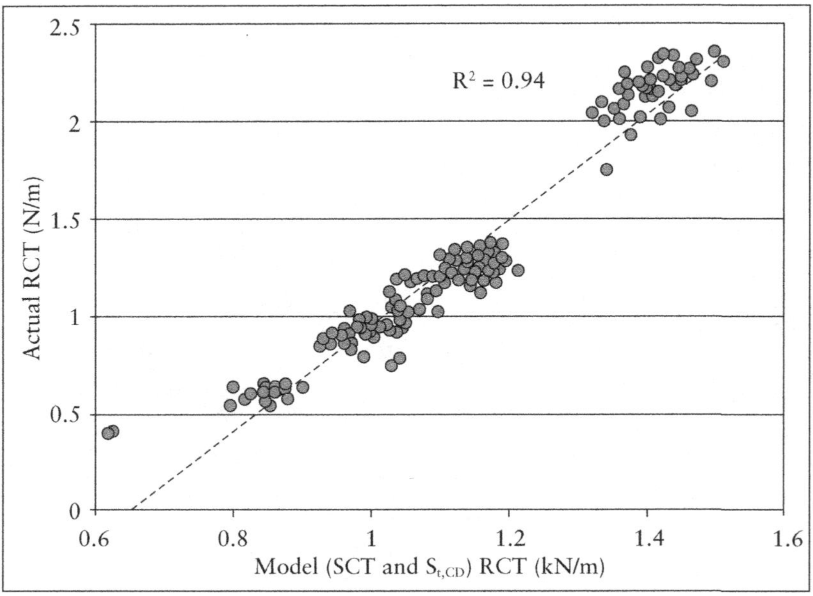

Figure 5.8 Actual RCT *versus* a fitted model of RCT using SCT and S_t for lightweight 100% recycled pulp linerboard

The fitted model for RCT represented in **Figure 5.8** was determined to be:

$$RCT = 0.42\ SCT^{0.86}\,(S_{t,CD})^{0.14} \tag{5.15}$$

with RCT, SCT and $S_{t,CD}$ all in units of N/mm. The average error between actual and model values is 0.27 N/mm. The advantage of such a model is that it uses the data available from automated testing machines avoiding the necessity of error-prone manual testing of RCT.

5.2 Compression Testing of Corrugated Board

The interest in testing SCT and RCT is primarily to optimise the compression strength of corrugated board. Shipping containers made from corrugated board are required to meet compression strength specifications in order to be certified. The compression strength of boards determines the stacking strength of boxes. The compression testing of boards is faster and easier than the compression testing of boxes hence, considerable attention is paid to the edge compression strength test (ECT) of board in box plants for production quality control. The effects of the quality of selected components and converting operations, such as the inadvertent crushing of board and bonding quality of the fluted medium to the linerboard, can be measured by ECT.

Similarly, for the compression testing of linerboard, consideration of the contribution of bending or other artefacts must be considered in order to obtain an optimal compression strength result. Details, such as specimen preparation, fixtures or constraints, have an effect. The board compression test of interest for most box-making operations is the edgewise compression in the CD ECT. In general terms, and to a good approximation, it is related to either SCT or RCT through a linear length-weighted summation of the individual strengths that make up the corrugated board [11]:

$$ECT = C\{(2\,SCT_l + \alpha\,SCT_m)\} \tag{5.16}$$

$$ECT = C'\{(2\,RCT_l + \alpha\,RCT_m)\} \tag{5.17}$$

where the medium to linerboard length ratio is the take-up factor 'α' and is typically 1.43 for C-flute boards. When using the RCT or

SCT summation it has been found that the form of the ECT equation assumes that the components all fail at equal deformation. The medium in most common boards is usually at a lower basis weight than the linerboards, so it will fail at a lower strain and strength when compression tested by itself alone; however, when it is fluted, the curvature imparts a structural strength such that the failure at equal deformation holds, as observations and numerical simulations show. The constant 'C' is less than 1, typically 0.7, reflecting the loss of strength of the components due to the converting process. The formula can be extended to multiwall boards by summing up the length-weighted SCT compression strengths of the '*i*' linerboards and '*j*' fluted media with their respective take-up factors:

$$ECT = C\left\{\sum_i SCT_i + \sum_j \alpha_j SCT_j\right\} \quad (5.18)$$

5.3 Three Different Edge Compression Strength Test Methods

When ECT values are substantially less than those predicted by the formula, it may be attributable to the method chosen not being suitable for the type of board being tested. Test specimen size, specimen preparation and compression strain rate all have an effect on the values. Three different methods are in common use. The traditional TAPPI T 811 method [12] hardens the vertically loaded edges of the test specimen by immersion of the edges in molten wax to a prescribed depth. A different height of the test specimens is cut, depending on the flute size, to limit bending, as may be assessed by calculation of the slenderness ratio and ensuring that it is less than 30. The vertical edges must be cut squarely and parallel to one another, and **Table 5.2** cites the various test specimen dimensions for various types of boards. Test pieces are supported vertically between the compression tester platens by machined steel blocks, which are removed from the test piece once contact is made by the advancing platens. The TAPPI T 811 method is approved for shipping box-stamp certification.

Table 5.2 Common flute sizes, TAPPI T 811 method test height, caliper and the slenderness ratio. 50 mm was selected arbitrarily as the test specimen height for E- and F-flutes

Flute size	TAPPI T 811 method test (height in mm)	Board caliper (mm)	Slenderness ratio (λ)
A	50.8	5.2	33.8
B	31.8	3.2	34.4
C	38.1	4.3	30.7
E	50	1.2	144
F	50	1	173

Reproduced with permission from R.E. Popil, *BioResources Journal*, 2012, 7, 2, 2553. ©2012, North Carolina State University [13]

The inconvenience of cutting different specimen heights for different board types and dipping edges in molten wax is removed in the TAPPI T 839 method [13], which uses a fixture consisting of two pairs of clamps that hold the test specimen vertically aligned with the compression platens. This is the method often used for quality control checks in box plants. The specimen dimensions remain constant for all board types and a specialised cutter, using parallel blades, ensures squareness and parallelism of the vertically loaded edges. A problem with the TAPPI T 839 method [9] can develop with high basis weight boards where rolling edges occur, or at lightweights, where the spring-loaded clamps introduce outward bowing of the facing of the test piece. When using the TAPPI T 839 method, the test pieces should be examined after compression to ensure that a visible crease, extending through the width of the specimen, is clearly visible on both sides of the board. **Figure 5.9** shows two commercially available TAPPI T 839 method ECT clamp fixtures. The springs provide a pressure of approximately 10 psi onto C-flute boards when the clamp is released to hold the board test piece firmly in place for the compression test.

a) b)

Figure 5.9 Sumitomo (a) and Emerson (b) TAPPI T 839 method ECT clamp fixtures

Figure 5.10 illustrates several types of failure that can occur with the clamp method. Heavy basis weight boards often display a folding compression of the edge at the top or bottom. The accepted test failure is observed as a crease going across the test piece at the unclamped middle portion. Failure is guaranteed to occur in the middle portion of the test piece once a tapered profile is cut into the test piece using the rotating knife fixture shown in **Figure 5.11**.

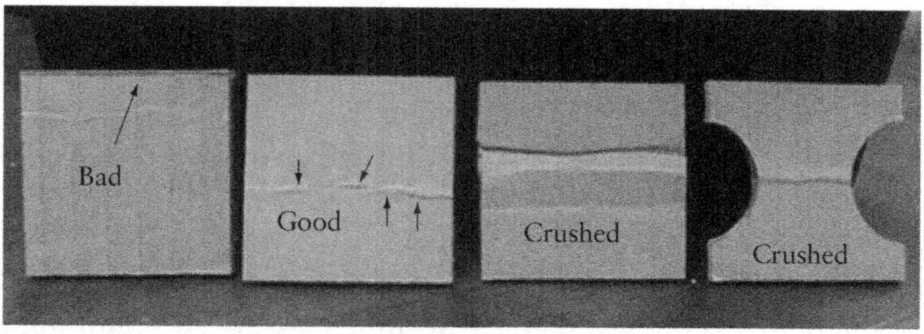

Figure 5.10 Examples of postECT tested board specimens. From the left, a sample with an MD edge roll, crease not at the midde; next, a good sample with a visible crease at the free unclamped middle of the test piece; a crushed board showing outward buckling failure; rightmost sample, a neckdown crushed sample still showing a crease at the middle of the test piece

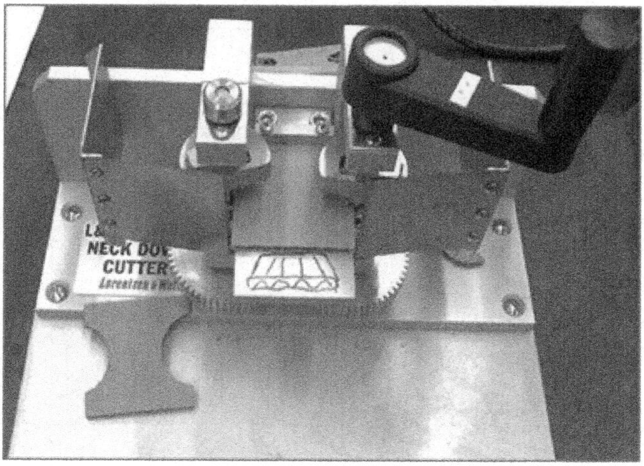

Figure 5.11 Fixture with rotating blades for cutting neckdown board pieces for the TAPPI T 838 method ECT

Figure 5.12 shows differences in ECT values using the different methods on a series of laboratory-made A-flute boards all comprising of 205 g/m^2 linerboards but with varying basis weight of fluted medium ranging from 68 to 205 g/m^2. The largest disagreement between predicted values and measurements occurs for the TAPPI T 839 method for low basis weight samples. The least disagreement overall for the sample set occurs for the TAPPI T 838 method. At high medium basis weights, the TAPPI T 839 method provides higher values than the TAPPI T 811 method, as may be expected since the board is under greater restraint when in the clamping test fixture. The large disagreement for the TAPPI T 839 method can be understood by the effects of the clamps crushing the sample. What appears to happen in the case of boards of low basis weight medium or boards, that have undergone crush damage, is that the unclamped facings of the test specimen have a small convex curvature at the onset of increasing load and the curvature increases to produce a noticeable bending failure rather than compression failure, as shown in **Figure 5.10** and **Figure 5.13**.

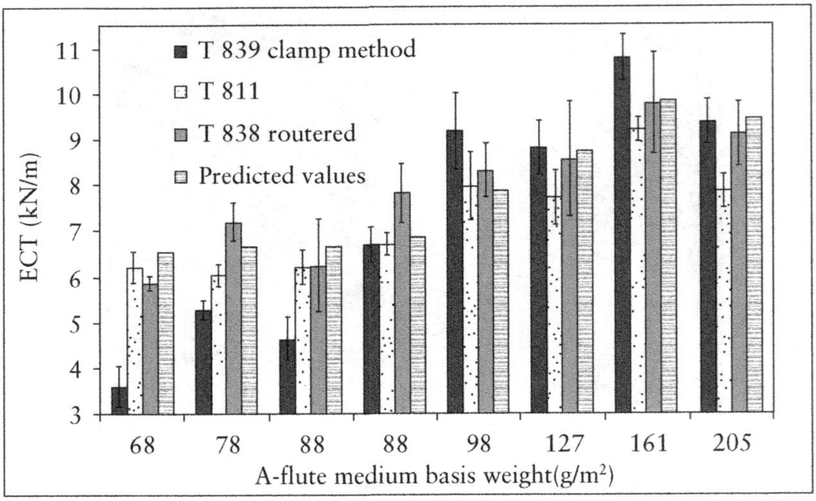

Figure 5.12 ECT values for A-flute boards made with 205 g/m^2 linerboards but different weights of medium as indicated. Reproduced with permission from R.E. Popil, *BioResources Journal*, 2012, 7, 2, 2553. ©2012, North Carolina State University [9]

Figure 5.13 The TAPPI T 839 method clamp test fixture with an insert showing a close up of the clamps crushing a low basis weight board. Reproduced with permission from R.E. Popil, *BioResources Journal*, 2012, 7, 2, 2553. ©2012, North Carolina State University [9]

Much corrugated board is crushed to some degree during manufacture either from the transfer press rolls or through die-cutting scoring and slitting operations that are not optimally maintained. A-fluted medium weakened from crushing can produce an anomalously low ECT value, similarly to the lightweight medium results shown in **Figure 5.9**.

In one study, a series of laboratory-made boards were run through a motorised laboratory roller press nip running at a speed of 50 ft/min with the roll gap separation set at various percentages of the original uncrushed caliper, ranging from 90 to 60%. A survey of box plants showed that a typical transfer roll gap is approximately 80% of the initial board caliper. In these investigations, boards had a fluting weight of 127 g/m^2 (26 lb/1,000 ft^2) and were of type A-, B- and C-flute, the final calipers after nip crushing were measured and the ECT results acquired using the TAPPI T 839 method. In addition, one series of C-flute boards with a lighter fluting weight of 112 g/m^2 (23 lb/1,000 ft^2) was also tested using both TAPPI T 839 and T 811 methods, and the results are summarised in **Figure 5.14** as the ECT loss as a function of caliper reduction. Generally, the crushing data indicates that ECT reduction is proportional to the corresponding decrease in caliper. The lightweight board 23C data (open circles and triangles in **Figure 5.14**) indicate that the waxed-end method results (triangular points) have lower ECT loss values compared with the clamp method results for low caliper reductions, as expected, but the situation reverses at high crush values.

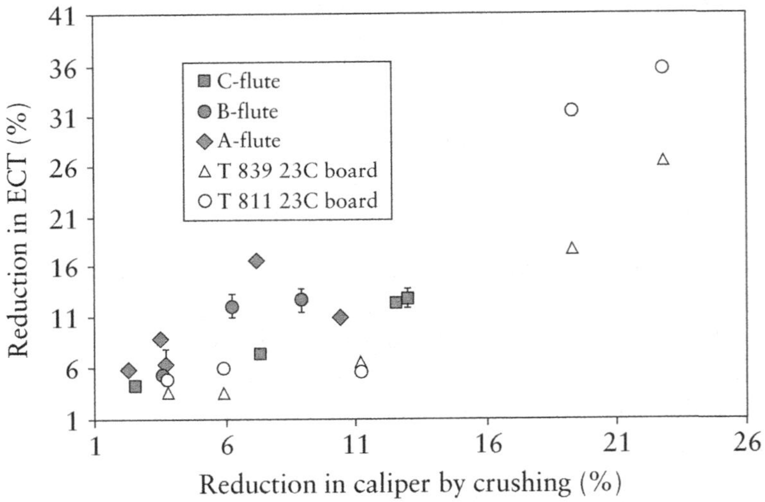

Figure 5.14 Reduction percentage in ECT value of boards progressively crushed by a rolling nip

5.4 Bending during Board Compression Testing

Bending during compression testing always lowers the value of the peak load. Bending of the whole test piece or, on a smaller scale, of the linerboard facings can occur. Bending of the entire test piece is controlled by limiting the test piece height, whereas the bending of the facings between the flute lines (interflute buckling) is governed by the facing stiffness and flute spacing. Corrugated board bending stiffness '*D*' is approximated as a sandwich structure consisting of facings S_b '$E \times t$' separated by caliper '*h*' such that the medium has a negligible contribution and thus:

$$D \cong \frac{Eth^2}{2} \tag{5.19}$$

The vertically loaded buckling load '*P*' of a corrugated board approximated here as a sandwich beam is considered to consist of a combination of beam buckling 'P_E' and shear 'P_S' [14], expressed as the harmonic mean in the form:

$$\frac{1}{P} = \frac{1}{P_E} + \frac{1}{P_S} \tag{5.20}$$

The Euler beam buckling load 'P_E' per unit length is π^2EI/H^2 or π^2D_b/H^2 with 'I' being the second moment of area, 'H' is the height of the vertically loaded beam and 'D_b' is the beam bending stiffness. The properties of relevance here are the elastic modulus (E_{CD}) and machine direction of paper bending stiffness (D_{CD}), as the vertical loading in the ECT is along the direction of the fluting or cross-direction, CD. If the ends of the beam are rigidly held or constrained from pivoting, then Euler beam buckling theory predicts the bending load to be $4\pi^2D_{CD}/H^2$.

Shear in corrugated board is the propensity of one facing to become displaced relative to the other when the board is subjected to stress.

Shear stiffness or shear rigidity is regarded as significant due to the following considerations. The shear rigidity, *Ps*, for corrugated board has been measured using a torsion pendulum method [15]. Essentially, this method involves using a long strip sample of board clamped at either end, with the upper end fixed to a supporting frame and the lower attached to a dumbbell that is free to rotate. The board is set to undergo a twisting motion as shown in **Figure 5.15**, the frequency of the twisting motion is converted to a shear rigidity using the D measurement and results from mechanical analysis.

Figure 5.15 Torsion pendulum measurement of the twisting frequency of a corrugated board strip in the MD

For typical C-flute board with medium basis weight of 127 g/m^2, shear rigidity in the CD, designated R_{44}, is 54 kN/m, which is much larger than the corresponding MD value of 9.8 kN/m. Buckling loads, determined by $\pi^2 D_b/H^2$ for typical C-flute samples having a D_{CD} of 5 N/m and height in the range of 50 to 32 mm, are calculated to be 20 to 48 kN/m, which are comparable to the measured shear rigidity.

Valid ECT measurements for any board must not have bending, which means that the critical buckling load '*P*' of a board must be greater that its compression strength ECT. Otherwise, as the compression platens advance and the load increases, the board will first fail by beam buckling rather than by compression. Upon substitution and rearranging the Plantema equation (**Equation 5.20**) this criterion, that $P >$ ECT, takes the form:

$$\frac{\pi^2 D_{CD} P_S}{\pi^2 D_{CD} + P_S H^2} \geq ECT \tag{5.21}$$

This equation is used for comparison with experimental data to assess the effect of test specimen height on ECT. This allows a check of whether the selected height is optimal for a particular board.

A selection of various corrugated board was chosen to evaluate the effects of height on ECT. The relevant properties of these boards is given in **Table 5.3**. D was measured using the four-point method and shear rigidity by the torsion pendulum method.

What is interesting to note in **Table 5.3** is how close the simple sandwich beam calculation for '*D*' is compared with the actual four-point method measured values and the more sophisticated formulas or calculations, such as those of Carlsson, Ranger or Nordstrand, which take into account the small contributon of the fluted medium.

Table 5.3 Relevant corrugated board properties used in the evaluation of test piece height on ECT

Board properties							D_{CD}					MD shear rigidity R_{44} (kN/m)
Board flute	Take-up factor α	Liner CD modulus E_l (GPa)	Medium CD modulus E_m (GPa)	Liner caliper t_l (mm)	Medium caliper t_m (mm)	Board caliper h (mm)	Four-point measured (N-m)	Sandwich beam formula (N-m)	Ranger formula (N-m)	Nordstrand formula (N-m)	Carlsson calc. (N-m)	
C	1.42	2.09	1.61	0.29	0.25	4.21	5.2	5.37	5.28	5.31	5.45	54.3
E	1.27	1.72	1.13	0.24	0.19	1.64	0.7	0.56	0.44	0.44	0.6	20
F	1.25	1.74	1.14	0.29	0.22	1.37	0.41	0.47	0.31	0.32	–	–
N	1.2	1.74	1.14	0.29	0.22	1.07	0.23	0.29	0.16	0.16	–	–

Reproduced with permission from R.E. Popil, *BioResources Journal*, 2012, 7, 2, 2553. ©2012, North Carolina State University [13]

5.5 Experimental Observations – Effect of Test Specimen Height

5.5.1 C-Flute Board

Two series of parallel experiments were conducted using commercially produced board WC 4226C 42 (205 g/m^2 liner and 126 g/m^2 medium), obtained off the corrugator, and similar laboratory-produced board using the same components. One series of test specimens had their ends embedded perpendicularly in a quick-setting epoxy resin, Alumilite™, producing a self-supporting test piece that is vertically compressed on its edge in a universal testing machine (UTM) compression tester, shown in **Figure 5.16**.

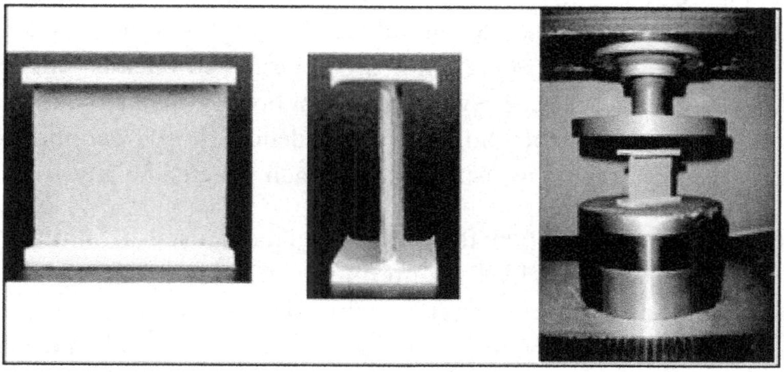

a) b) c)

Figure 5.16 a) and b) an ECT sample embedded in supporting epoxy resin strips, front and side views, and c) mounted between the platens of a compression tester. Reproduced with permission from R.E. Popil, *BioResources Journal*, 2012, 7, 2, 2553. ©2012, North Carolina State University [9]

This is a variation of the TAPPI T 811 method with the horizontal test piece ends encapsulated in hardened epoxy resin platforms instead of being dipped in molten wax. The test pieces were prepared using fixtures that supported the cut boards vertically while the setting epoxy was poured. The epoxy platforms restrict pivoting at the ends and ensure failure will occur away from the edges in the same manner as the impregnated hardened wax. Test pieces were compression tested in a Model 1122 Instron UTM using Series IX software to record load and deformation at a cross-head velocity of 12.5 mm/min.

Figure 5.17 shows the experimental data of the resin-embedded ends of the commercial WC 4226C C-flute board blank (205 g/m^2 liner, 127 g/m^2 medium single-wall) sample set. The data set suggests that the free span between the glued ends, i.e., a specimen height of up to 80 mm, will provide a constant value of ECT. The fall in ECT value attributable to the onset of beam bending is in general agreement with the estimate provided by the beam buckling criterion equation, which is in **Figure 5.17**, and is used to plot height *versus* buckling load using the board properties detailed in **Table 5.3**. Boards that have an ECT at a height above the curve value are expected to buckle, whereas ECT values under the curve value represent boards that are failing *via* compression with no bending. Error bars denote the 95% confidence intervals of 10 repeat measurements at each selected height.

An alternative specimen testing arrangement used a Sumitomo TAPPI T 839 method ECT clamp fixture, which was disassembled to accommodate differing sets of machined aluminium block spacers to separate the upper and lower clamps from the default standard setting of 11.2 to 110 mm. Similar results for various heights are shown in **Figure 5.18** using the same commercially produced C-flute board used for **Figure 5.17**.

The data of **Figures 5.17** and **5.18** indicate that a WC 4226C test piece height of 60 mm or less ensures an ECT will not be influenced by beam buckling provided the ends are embedded in resin platforms or are held in the clamp fixture. Throughout this discussion, 'free span' is the height of the board that is free of epoxy resin or is not in contact with the jaw faces of the TAPPI T 839 method fixture.

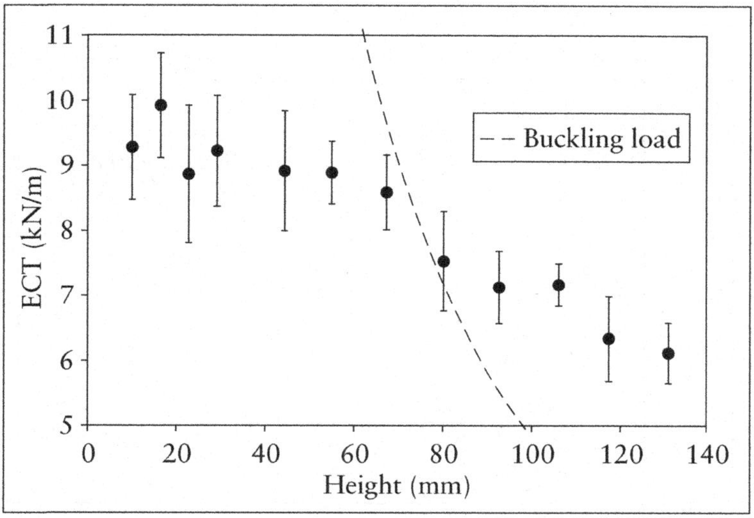

Figure 5.17 ECT *versus* the span length for C-flute specimens with the horizontal edges embedded in resin. The dashed curve is the calculated buckling load which indicates the height at which beam bending is expected to dominate the failure mode. Reproduced with permission from R.E. Popil, *BioResources Journal*, 2012, 7, 2, 2553. ©2012, North Carolina State University [9]

The error bars in the clamped method data shown in **Figure 5.17** are somewhat higher than the resin-embedded end method results shown in **Figure 5.18**. This is probably partly attributable to some degree of compression of the edges during the clamp method *versus* resin-embedded ends, and possibly some lack of true parallelism of the clamps with the progressive insertion of machined block assemblies to increase the spacing between the clamps. The expected ECT for this board from calculation is 8.9 kN/m, which is consistent with measurements for spans less than 60 mm using both methods.

Observations of lower than expected values of ECT at very short heights, reported by McKee in 1961, were probably an artefact.

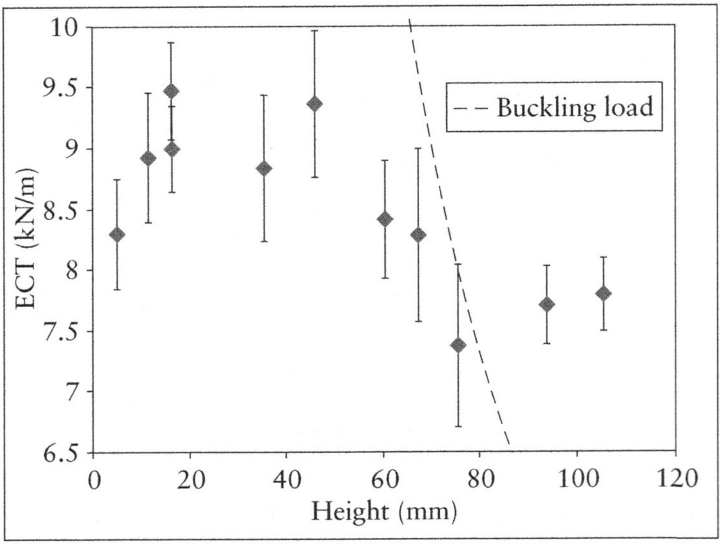

Figure 5.18 ECT as a function of the free span height for C-flute specimens using a modified Sumitomo clamp to extend the test specimen unclamped height to 110 mm. The dashed curve is the buckling load using the measured D and shear stiffness. Reproduced with permission from R.E. Popil, *BioResources Journal*, 2012, 7, 2, 2553. ©2012, North Carolina State University [9]

5.5.2 E-, F- and N-Flute Boards

Compression strength measurements for smaller flute boards allow calculation of the potential stacking strength for storage or transport applications. Miniflute boards are replacing the use of folding cartons in many instances. There are no accepted standards for compression measurement in this case, so it is useful to qualify a current ECT method for miniflutes. Samples of commercially made E-, F- and N-flute board blanks were cut to various lengths and tested in a similar fashion to that described above for the C-flute board. These boards all had a liner of 207 g/m^2 and fluted medium of basis weight 112 g/m^2.

The results for resin-embedded edges of E-flute specimens are shown in **Figure 5.19**. A fall in ECT occurs with increasing free span height, as expected from increased bending occurring at larger heights. Calculations for the buckling-load curve in **Figure 5.19** are the same as for C-flute but using D_{CD} and P_S values for E-flute from **Table 5.3**.

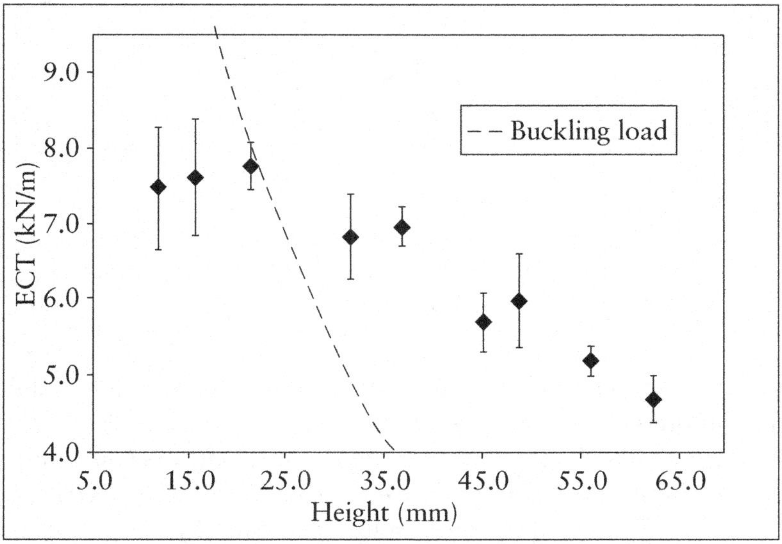

Figure 5.19 ECT as a function of the free span length for resin-embedded edges of E-flute test specimens. Reproduced with permission from R.E. Popil, *BioResources Journal*, 2012, 7, 2, 2553. ©2012, North Carolina State University [9]

Figure 5.20 shows the ECT *versus* free span height for Sumitomo clamped E-flute board. **Figures 5.19** and **5.20** suggest free spans of 25 mm or less will suffice to provide a representative compression strength value for E-flute. The default free span for a 50.8 mm Billerud cut square test piece in the TAPPI T 839 method is 11.2 mm; therefore, the TAPPI T 839 method can be expected to provide a reliably accurate value for this single-wall E-flute board.

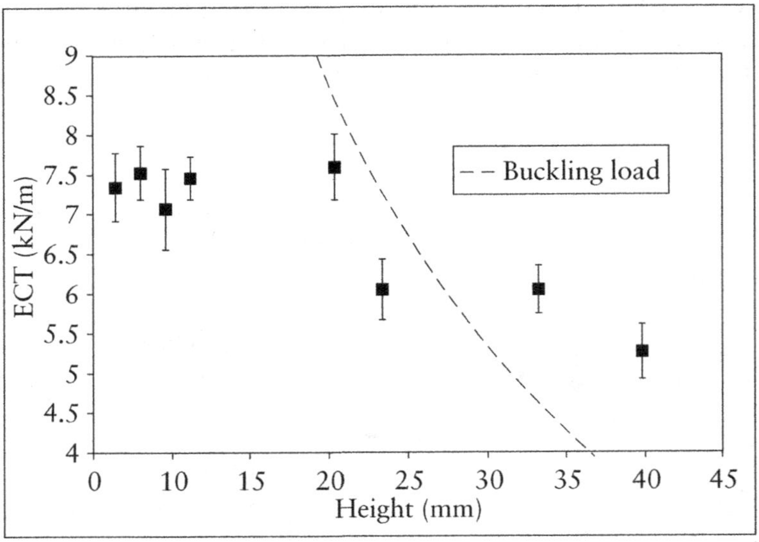

Figure 5.20 E-flute specimens cut to various lengths and tested in a Sumitomo clamp modified with extensions (dashed curve is the calculated buckling load). Reproduced with permission from R.E. Popil, *BioResources Journal*, 2012, 7, 2, 2553. ©2012, North Carolina State University [9]

A similar study was conducted employing 42-23F and 42-23N microflute boards using the TAPPI T 839 method clamp only and set to provide various free spans. The data shown in **Figure 5.21** display a plateau in ECT at a height of 9 mm or less.

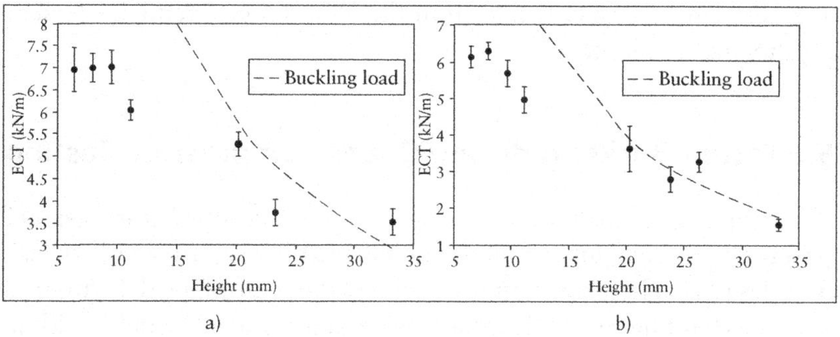

Figure 5.21 ECT *versus* free span height using the TAPPI T 839 method clamping fixture for 42-23F (a) and 42-23N (b) board. Dashed curves represent calculated vertical buckling loads. Reproduced with permission from R.E. Popil, *BioResources Journal*, 2012, 7, 2, 2553. ©2012, North Carolina State University [9]

The plateau regions of constant ECT values in **Figure 5.21** appear to be overestimated by the buckling condition, i.e., P > ECT, **Equation 5.21**, with the shear rigidity for E-flute as an approximation for F- and N-flutes. **Figure 5.21** indicates that the TAPPI T 839 clamp method can be used for F- and N-flutes if the free span does not exceed 9 mm. This can be easily achieved by cutting the microflute samples to a height of 47 mm or smaller prior to their insertion into the TAPPI T 839 method clamping fixture.

The data presented here augment and substantiate previous observations of the effect of specimen height on ECT [16–20]. ECT values will decrease with increasing specimen height due to specimen buckling rather than specimen compression. Specimen heights for optimal ECT measurement fall approximately within the limits estimated by sandwich structure buckling. The implication is that the beam mechanics shown here can be applied to other board configurations to estimate an appropriate specimen test height once a measurement or a calculation of the D and measurement of shear rigidity are available.

5.6 Facing Buckling during Board Compression Testing

If corrugated boards undergoing a vertical compression test are observed closely, quite often the facings exhibit a pattern of surface dimples [21]. These are actually sinusoidal out-of-plane deformation waves extending along the glue lines resulting from panel buckling of the facings [22] due to the applied vertical load. Whenever the compression strength is greater than the load required for buckling, i.e., the critical buckling load, the linerboard facing will buckle in a wave pattern. To a good approximation, the critical buckling load for the facings are given by the formula for a simply supported panel of width 'w' and MD and CD bending stiffnesses 'D_1' and 'D_2':

$$P_{cr} = \frac{4\pi^2 \sqrt{D_1 D_2}}{Kw^2} \tag{5.22}$$

'K' is a factor to account for the constraint of the plates, which = 1 for the case of a simple support, i.e., the edges are free to rotate. So, the criterion for linerboard buckling to occur stated more specifically is that the CD SCT must be greater than 'P_{cr}', much the same criterion previously discussed for boards not to have bending expressed in **Equation 5.21**. The critical buckling load for the linerboard facing is calculated using the flute glue line spacing, e.g., it is 8 mm for C-flute and the geometric mean of the bending stiffness $\sqrt{(D_1 D_2)}$. The constant K in P_{cr} accounts for whether the panels are simply (= 1) or rigidly supported (= 2).

D is measured using the Taber instrument of L&W with two-point bending measurement. When using the Taber instrument, recall that the instrument read-out values must be multiplied by 0.164 to convert the read-out of bending moment of grams-force to D in mN-m.

A patterned dimpling of linerboard facings is commonly observed in corrugated fibreboard containers that have been placed under a vertical-stacking load and exposed to a high-humidity environment. The same dimpling or localised buckling is also observed to occur during the ECT of large-flute (A or C) lightweight corrugated boards. A time sequence of video recordings of the ECT of various boards were made with the frames synchronised with load-displacement data. An example of a few selected frames from one such sequence is shown in **Figure 5.22**, which tested a WC 3526C board mounted in supporting resin platforms.

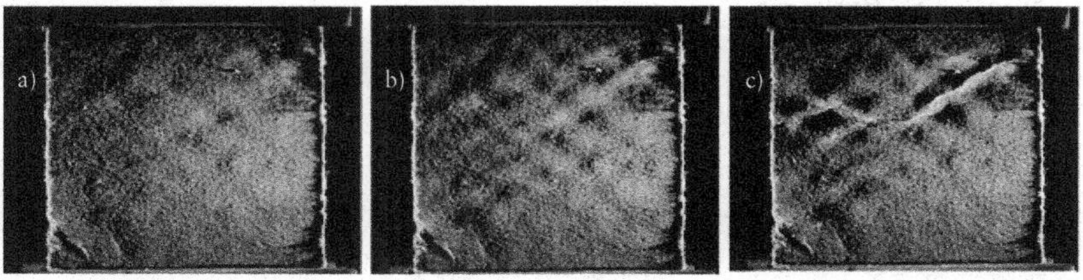

Figure 5.22 Excerpted time sequence video snapshots of a glancing angle illuminated ECT test of a WC 3526C-flute board: a) 5.7 kN/m load, 1.3% strain, some buckling is evident, b) peak load 6.7 kN/m, 1.6% and c) postfailure 4.5 kN/m, 2.1%, crease forms joining microplate buckling crests. Reproduced with permission from R.E. Popil, *BioResources Journal*, 2012, 7, 2, 2553. ©2012, North Carolina State University [9]

Glancing angle collimated illumination from the right-hand side in **Figure 5.22** was used to highlight features on the linerboard. Visible dimpling of the facing is observed to progressively form as the load increases near to the peak value. Once the peak load is reached, a crease immediately forms which progressively develops joining the crests of the buckled areas across the width of the sample. For boards with a small flute size (B or E) or heavyweight linerboard facings, no buckling patterns are observed, only the formation of a crease at peak load.

The fluted medium is assumed not to buckle due to the additional stiffness imparted by its curvature. Numerical simulations of ECT, using the material constants for the laboratory-made boards used in a study at the author's laboratory, were made using a non-linear constitutive model geometric non-linear effects [23]. Patterned out-of-plane dimpling occurs in the simulations, which corresponds to experimental video observations. The magnitude of the out-of-plane displacement of the buckling pattern of the 205 g/m^2 linerboard facings is as much as 0.3 mm in the simulations. The computations indicate no corresponding buckling for the 127 g/m^2 medium, as indicated in **Figure 5.23**.

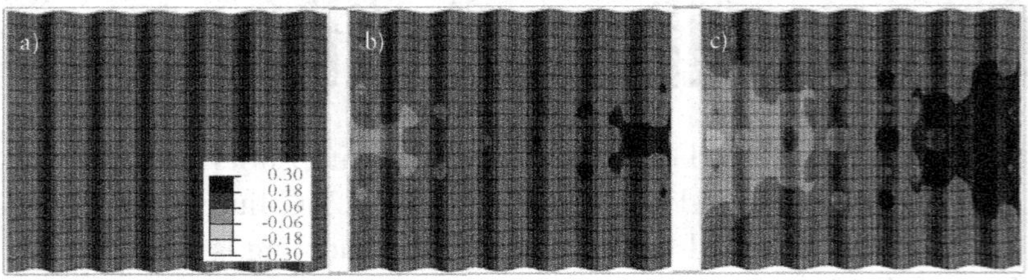

Figure 5.23 Progressive out-of-plane displacement maps, a) to c), for the C-flute medium from a non-linear numeric finite element analysis (FEA) ECT simulation by Haj-Ali and co-workers. Reproduced with permission from R.E. Popil, *BioResources Journal*, 2012, 7, 2, 2553. ©2012, North Carolina State University [9]

Figure 5.23 shows the progressive calculated out-of-plane displacements for the medium for C-flute simulation data presented in [23]. The numeric results indicate that the fluted medium does not buckle like the facings during a compression test. In **Figure 5.23a**, the failure initiates with the outer facing (not shown in **Figure 5.23**) starting to show a patterned buckling and a Tsai-Wu failure criterion (a summation of directional stresses exceeding yield stresses) is met locally corresponding to the start of a formation of a crease, however, the medium (**Figure 5.23a**) shows no patterned buckling. **Figure 5.23b** is the calculated out-of-plane displacement at the ultimate failure stress level, which is determined by the Tsai-Wu criterion linking several failed regions across the width of the board. At this point in the simulation, the FEA model of Haj-Ali and co-workers [23] shows a prevalent predicted patterned buckling for the facings at failure resembling **Figure 5.22b** and **Figure 5.22c**, however, the medium in **Figure 5.23** shows negligible buckling. A third calculation result is shown in **Figure 5.23c** postfailure, where the vertical displacement is 2X the failure initiation displacement.

Video recordings were also made of laboratory-prepared A-flute 127 g/m^2 test pieces adhered to clear plastic film and subjected to vertical load to see if any fluted medium buckling occurred. In this case, calculations indicate that the buckling load for the flat portion of the fluting is 12.6 kN/m and for the curved portion, of radius 1.5 mm, the calculated buckling load is 209 kN/m. These calculated buckling load values are much larger than typical single-wall A- and C-flute ECT values. Indeed, the video recordings of A-fluting adhered to plastic film test pieces showed no buckling prior to a compression crease failure, which always coincided with plastic film creasing.

Hence, video observations, and analytical and numerical simulations all indicate that the 127 g/m^2 medium will not buckle in ECT, but the linerboard facings of 205 g/m^2 will buckle, provided that their compression strength exceeds their panel buckling load for a given flute spacing. If the linerboard facings buckle, it is expected that their ultimate strength will be governed by the empirical expression for plate failure load, 'P_z':

$$P_z = cP_m^b P_{cr}^{1-b} \tag{5.23}$$

where, per unit length, 'P_m' is the intrinsic compression strength of the material, 'P_{cr}' is the critical buckling load given by **Equation 5.22**, and '*c*' and '*b*' are empirical constants. **Equation 5.23** is the basis for the derivation of the McKee equation for the compression strength of boxes consisting of linked buckling panels with P_m being the ECT of the corrugated board. By similar reasoning, ECT consists of buckling linerboard plates for which the appropriate P_m is now taken as the SCT. Thus, for a single-wall board, the model for ECT, which is considered as the equivalent of P_z, is written as:

$$\begin{aligned} ECT &= C'\{2\times(SCT)_l^b(P_{cr})^{(1-b)} + \alpha(SCT_m)\}\ \textit{if}\ (SCT)_l \geq P_{cr},\ \textit{buckling occurs} \\ &= C\{2\times(SCT)_l + \alpha(SCT)_m\}\ \textit{if}\ P_{cr} \geq (SCT)_l,\ \textit{no buckling} \end{aligned} \tag{5.24}$$

Application of **Equation 5.24** to multiwall boards is straightforward by the addition of similar terms to account for additional board components and different medium take-up factors α.

A large series of commercial and laboratory-made corrugated boards were considered for their buckling potential, calculated simply as the ratio of the buckling load of the interflute linerboard panel to the compression strength. In **Table 5.4**, the relevant physical properties of the components were measured: caliper, basis weight, CD SCT, MD and CD Taber stiffnesses.

Table 5.4 Summary of various corrugated board properties used in the study of buckling effects on ECT. The last column is P_{cr}/SCT which when ≤1 indicates possible linerboard buckling

Board sample ID	Board description	Board caliper (µm)	Basis weight (g/m^2)	SCT_{CD} liner (kN/m)	Liner density (g/cm^3)	SCT_{CD} medium (kN/m)	Taber RMS moment (gf-cm)	TAPPI T 839 ECT (kN/m)	Buckling load/ SCT
WC 3526C 35	Commercial C-flute	4,140	542	3.9	0.62	2.1	25.4	7.5	0.72
WC 3526C 35 IPST	Same as above but lab made	4,100	583	3.9	0.62	2.1	25.4	8.4	0.72
WC 3526E 35	Commercial E-flute	1,620	521	3.9	0.69	2.1	25.4	8.1	3.68
WC 4226C 42	Commercial C-flute	4,210	605	3.4	0.68	2.1	31.4	7.9	1.02
WC 4226C 42 IPST	Same as above but lab made	4,130	646	3.4	0.69	2.1	31.4	8.7	1.02
WC 5626C 56	Commercial board	4,350	748	5.3	0.84	2.1	65.7	11.2	1.34
WC 5626C 56 IPST	Same as above but lab made	4,290	757	5.3	0.80	2.2	62.6	11.0	1.28
GP 262C 26	Commercial C-flute	3,838	420	2.4	0.68	2.1	8.5	5.3	0.38
GP 3526C 35	Commercial C-flute	3,957	507	3.7	0.66	1.9	19.1	6.8	0.56
GP 4233C 42	Commercial C-flute	4,045	575	3.9	0.59	2.2	31.1	7.9	0.87
GP 5535C 55	Commercial C-flute	4,334	783	5.3	0.80	2.2	62.6	11.0	1.28
TI 4223E 42	Commercial E-flute	1,691	581	3.8	0.72	1.7	27.0	8.0	3.99
TI 3523C 35	Commercial C-flute	4,057	547	3.7	0.70	2.4	20.2	7.8	0.60
TI 5623C 56	Commercial C-flute	4,207	751	5.4	0.70	1.8	63.8	10.8	1.29
IPST 4226C 35	Asymmetric lab board	4,214	580	4.6	0.70	2.4	42.2	8.6	0.95
IPST 4226C 55	Asymmetric lab board	4,443	687	4.6	0.70	2.4	42.2	10.6	0.95
IPST 4226E 4226C 42	Lab-made multiwall board	5,679	1,019	4.6	0.70	2.4	42.2	16.3	0.95

IPST 4226A 42	Lab-made A-flute	5,171	624	4.6	0.70	2.4	42.2	8.5	0.81
IPST 4226B 42	Lab-made B-flute	3,044	599	4.6	0.70	2.4	42.2	9.9	1.64
IPST 4226A 4226B 42	Lab A/B-flute multiwall	7,933	1014	4.6	0.70	2.4	42.2	14.9	0.81
IPST 4226A 4226C 42	Lab A/C-flute multiwall	9,015	1011	4.6	0.70	2.4	42.2	14.9	0.81
IPST 4226B 4226C 42	Lab B/C-flute multiwall	6,959	1026	4.6	0.70	2.4	42.2	15.3	1.64
IPST 4226C 42	Lab-made C-flute	4,252	613	4.6	0.69	2.4	42.2	9.5	0.96
IPST 4226C 42	Lab-made C-flute 2nd set	4,215	588	4.6	0.69	2.4	42.2	9.6	0.96
IPST 4226E 42 trial	Lab-made E-flute	1,660	582	4.1	0.70	2.3	33.4	7.9	4.58
IPST 4226E 42	Lab-made E-flute 2nd set	1,642	591	4.6	0.70	2.4	42.2	10.5	5.16
N&W 100 low press	Lab-spliced handsheet	4,063	421	1.8	0.40	2.4	20.4	4.7	1.18
N&W 100 med press	Lab-spliced handsheet	3,958	420	2.1	0.52	2.4	11.4	4.5	0.56
N&W 100 high press	Lab-spliced handsheet	3,956	427	2.6	0.62	2.4	2.1	4.7	0.09
N&W 200 low press	Lab-spliced handsheet	4,352	634	3.6	0.41	2.4	78.2	7.6	2.27
N&W 200 med press	Lab-spliced handsheet	4,109	640	5.1	0.58	2.4	43.0	8.6	0.87
N&W 200 high press	Lab-spliced handsheet	4,092	629	5.9	0.71	2.4	32.3	8.4	0.57
N&W 300 low press	Lab-spliced handsheet	4,700	845	4.5	0.41	2.4	228.6	9.2	5.32

N&W 300 med press	Lab-spliced handsheet	4,357	871	6.6	0.56	2.4	161.8	10.5	2.54
N&W 300 high press	Lab-spliced handsheet	4,324	895	7.7	0.64	2.4	152.3	11.1	2.06
N&W 100 100 psi	Lab-spliced handsheet	4,357	390	1.5	0.44	2.4	8.9	4.5	0.61
N&W 100 400 psi	Lab-spliced handsheet	4,324	390	2.0	0.61	2.4	4.9	4.4	0.26
N&W 100 1,900 psi	Lab-spliced handsheet	4,507	390	2.2	0.80	2.4	2.3	4.5	0.11
N&W 160 100 psi	Lab-spliced handsheet	4,179	527	2.8	0.48	2.4	35.9	6.1	1.35
N&W 160 400 psi	Lab-spliced handsheet	4,147	530	3.5	0.64	2.4	27.0	6.2	0.79
N&W 160 1,900 psi	Lab-spliced handsheet	4,025	535	4.0	0.85	2.4	13.3	7.0	0.34
Formette 100 0 psi	Lab-spliced handsheet	4,105	410	1.1	0.29	2.4	17.4	3.9	1.69
Formette 100 50 psi	Lab-spliced handsheet	4,068	412	1.5	0.51	2.4	6.2	4.3	0.44
Formette 200 0 psi	Lab-spliced handsheet	4,263	638	2.2	0.30	2.4	104.0	6.6	4.82
Formette 200 50 psi	Lab-spliced handsheet	4,145	645	3.8	0.58	2.4	49.7	6.9	1.38

N&W: Noble and Woods

Reproduced with permission from R.E. Popil, *BioResources Journal*, 2012, 7, 2, 2553. ©2012, North Carolina State University [13]

Table 5.4 is a summary of various corrugated board properties used in the study of buckling effects on ECT. The last column is P_{cr}/SCT, which when ≤1 indicates possible linerboard buckling.

One series of boards produced on a pilot single-facer corrugator contained embedded-spliced handsheet linerboards of various basis weights pressed to various densities. These are designated by the N&W prefix and the Formette prefix in the sample ID in **Table 5.4.** This is similar to the investigation previously reported by Whitsitt in 1988, where linerboard handsheets of various basis weights were pressed to different densities to obtain a range of increasing compression strengths. Generally, increasing sheet density through wet pressing will increase the compression strength while decreasing the caliper. The furnish composition of the linerboard and medium were constant throughout the series of prepared handsheets. The prepared linerboard handsheets were individually manually spliced into conventional linerboard rolls that were run through the pilot single-facer corrugator at IPST using conventional corrugating operating parameters. Double-backing of the prepared single-face combined board samples was performed manually using a metered rolling nip, applying Stein–Hall starch adhesive to the flute tips of single-facing samples, followed by a hot plate press adhering linerboard sheets to the glued single-wall.

A second series of boards, designated by the IPST prefix or suffix in the sample ID in **Table 5.4**, was prepared using commercial linerboard and medium of various flute sizes and combinations utilising the IPST pilot-plant facilities, as described in detail by Schaepe and Popil [24]. In this second series, the linerboard and medium were the same throughout and exhibited the properties of the C-flute board data shown in **Table 5.4.** The ratio of buckling load P_{cr} to SCT shown in the last column in **Table 5.4** shows that many of the boards investigated could be expected to display interflute buckling whenever the ratio is less than 1. For the case of multiwall board samples, a single-ratio value is reported in the table column for the facing expected to buckle. Video recordings of the ECT testing of these boards confirmed the prediction of P_{cr} **Equation 5.22** with K

set to 1 for buckling onset. Specifically, **Equation 5.22** predicts that A- and C-flute boards will buckle when placed under vertical load, but B- and E-flute boards will not. The smaller flute size, '*w*', for B- and E-flute board makes the ratio P_{cr}/SCT > 1 and so no buckling is expected. Video recordings of the ECT of multiwall boards of both single-face and double-back sides simultaneously [21] showed the separate development of interflute buckling or compression failure depending on the flute size adhered to the specific facing.

A third series of boards selected for ECT buckling investigations consisted of commercial linerboards and medium supplied directly from the corrugator as unconverted cut sheets. The linerboard and medium for these boards were provided separately for their physical property measurements.

The analysis is divided into the types or classes of boards. In each case, a multiparameter fit for the constants *C*, *C*', *K* and *b* was applied to the sets of data using the model represented by **Equation 5.24.** Fitting was performed by iteratively reducing the error between the model and actual data using numerical routines such as the Solver algorithm in MS Excel.

Figure 5.23 shows the comparison of predicted ECT values from the model to actual ECT values for the series of laboratory-made boards using spliced handsheets. In cases where buckling of the facings occurs, the model places the predicted adjusted values closer to the identity line, reducing the error. Constants *C*', *C*, *b* and *K* were determined to be 0.72, 0.7, 0.65 and 0.96, respectively. *K* being close to 1 supports the model of linerboard plates between glue lines being simply supported at the unloaded vertical edges of the glue lines. The regression r^2 value increases from 0.91 to 0.94 when the buckling model, **Equation 5.24,** is used instead of **Equation 5.18,** and the mean error correspondingly reduces from 0.19 to 0.16 kN/m.

Buckling of the liners limits the compression strength of the corrugated board. Papermakers strive to produce liners with the highest possible compression strength at the lowest possible basis weight. Large amounts of refining energy and high-pressure wet pressing produces sheets of high strength but at a lower caliper, which will result in a lower D, although the elastic modulus will have increased due to the increased sheet density. Boards prepared and tested to produce **Figure 5.24** using spliced handsheets were examined for the relationship of caliper, compression strength SCT and D as a result of increasing wet pressing.

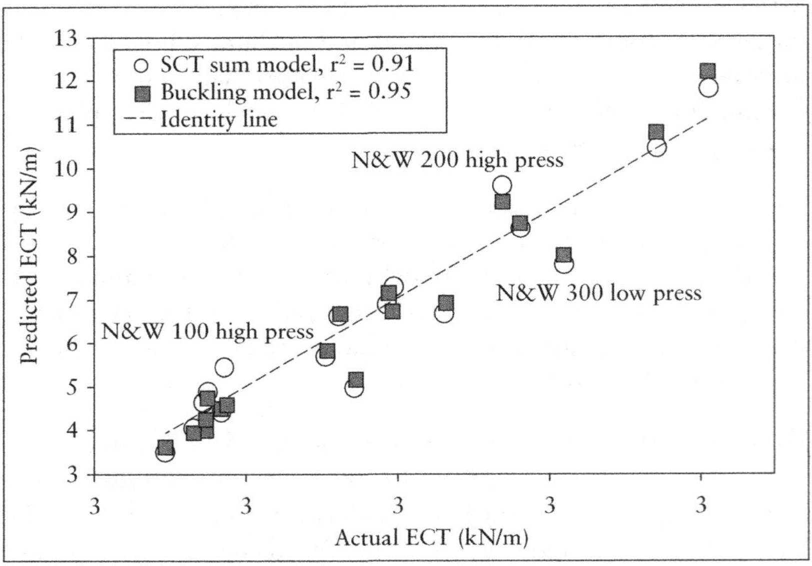

Figure 5.24 Comparison of the buckling model *versus* actual ECT values for a set of boards made with spliced handsheets pressed to high densities. Reproduced with permission from R.E. Popil, *BioResources Journal*, 2012, 7, 2, 2553. ©2012, North Carolina State University [9]

The controlled wet pressing of handsheets allowed a large range of densities to be prepared such that relationships between the SCT, $\sqrt{D_1 D_2}$ and basis weight could be determined, as shown in **Figure 5.25**.

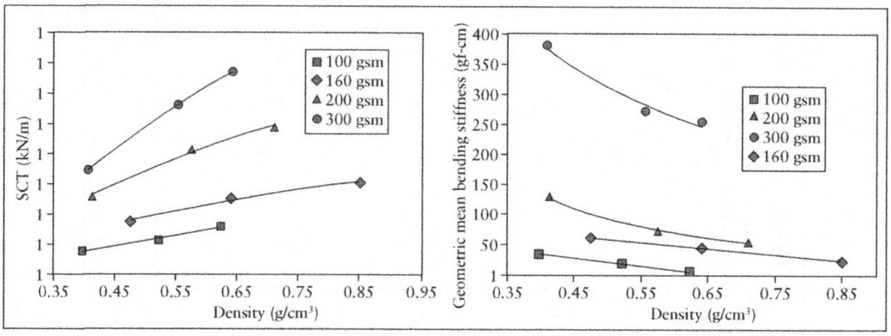

Figure 5.25 Relationships between SCT (left) and geometric mean, D (right), *versus* density for the handsheets used in the corrugated board test results shown in **Figure 5.24**

Previously shown by Whitsitt [25], and evident in **Figure 5.25**, increasing the compression strength SCT through wet pressing decreases the D and ultimately places a limit on the ECT that may be gained through wet pressing. Increased wet pressing results in higher density and hence SCT increases, in addition, from first principles, more fibre bonding may be expected [2]; however, the S_b simultaneously decreases as a result of reduced caliper. Therefore, the increase in SCT due to a wet pressing density increase is compromised by a corresponding decrease in D and this phenomenon is accounted for in the buckling model (**Equation 5.24**) for ECT. The respective values for SCT and $\sqrt{D_1 D_2}$ for each basis weight sample set (100, 160, 200 and 300 g/m^2) was substituted into the buckling model with fitted constants C and b to produce a series of predicted ECT values for each basis weight class as a function of the wet pressed density. **Figure 5.26** summarises these calculations which show optimal linerboard densities are reached for maximum ECT in lightweight linerboards (<200 g/m^2) adhered to 126 g/m^2 C-flute medium.

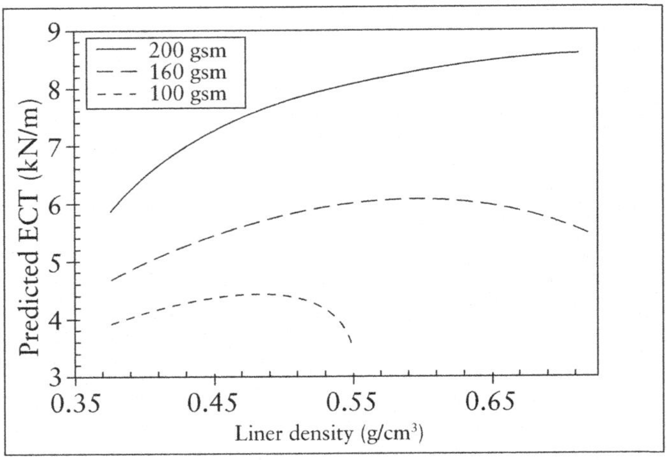

Figure 5.26 Summary of calculations for predicted ECT using the buckling model (**Equation 5.24**) and fitted constants derived from the handsheet data set of **Figure 5.24**

Figure 5.26 shows how once a certain density is exceeded, gains in SCT become offset by losses in D with the result that ECT does not increase. Thus, there is a limit to the benefits of wet pressing to increase SCT as, when incorporated into the corrugated board structure, the increase in compression strength becomes offset by the decrease in buckling load.

Application of the buckling model to predict ECT was also applied separately to the laboratory-made board sets (IPST prefix in **Table 5.2**), as well as the commercially made board set of **Table 5.2**. Whenever SCT > P_{cr}, the predicted ECT values were calculated using the buckling model, otherwise the linear summing SCT calculation was made. The summary of the analyses listed in **Table 5.5** shows that in all three sets of boards investigated, the overall error in ECT prediction [mean square-root error (MSE), last column] is reduced and the correlation with actual values improves.

Table 5.5 Summary of SCT sum and buckling model fits for several corrugated board data sets. MSE is the mean square-root error for each fitted model

Data set	ECT model	C or C'	b	K	r^2	MSE (kN/m)
All IPST lab-made board	SCT sum	0.70	–	–	0.91	0.19
	Buckling	0.72	0.65	0.96	0.94	0.16
Commercial boards	SCT sum	0.77	–	–	0.90	0.18
	Buckling	0.80	0.71	1.16	0.97	0.09
IPST-spliced handsheet boards	SCT sum	0.70	–	–	0.91	0.16
	Buckling	0.65	0.85	1.18	0.95	0.14

Reproduced with permission from R.E. Popil, *BioResources Journal*, 2012, 7, 2, 2553. ©2012, North Carolina State University [13]

5.7 Compression Strength of Boxes – McKee Formula

The compression strength of boxes [26] is considered to be the combined compression and buckling of four joined panels so that the general formula applies again:

$$\frac{P_{\max}}{P_{cr}} = \alpha\left(\frac{P_{com}}{P_{cr}}\right)^{b} \tag{5.25}$$

Such that P_{max} here for a panel of length 'L' is the box compression test strength (BCT)/L, P_{com} is the ECT (kN/m) of the corrugated board and P_{cr} is the buckling load (N-m) of the panels of the box given by:

$$P_{cr} = \frac{4\pi\sqrt{(D_1 D_2)}}{L^2} \tag{5.26}$$

So that the BCT becomes:

$$BCT = C\ ECT^{b}\left(\frac{\sqrt{D_1 D_2}}{L^2}\right)^{1-b} L \tag{5.27}$$

The box is assumed to be square such that its perimeter '*Z*' is simply 4*L*. The exponent '*b*' turns out to be ¾ so that the compression strength formula takes the form:

$$BCT = C\ ECT^{\frac{3}{4}}(\sqrt{D_1 D_2})^{\frac{1}{4}}\sqrt{Z} \tag{5.28}$$

A further simplification occurs with the assumption that ECT is proportional to the S_b of the liners $E_{1,2} \times t$, '*h*' being the board caliper, using the sandwich approximation for the board *D*:

$$D_{1,2} \cong \frac{E_{1,2}}{2} th^2 \tag{5.29}$$

The simplified form of the McKee equation as used in trademarked CAPE and TOPS computer-aided design models (commercially available box and pallet design software packages) is:

$$BCT = C'ECT \times \sqrt{h \times Z} \tag{5.30}$$

The assumption and substitution of D_1 and D_2 by ECT, thus removing the power dependence of BCT on ECT, ignores the contribution of the medium to ECT. Thus, predictions that are more accurate can be made using the formula using both ECT and *D*.

An example of improved accuracy in predicting BCT is presented for 15 different commercially prepared boards provided by the same box plant. The samples represent a variety of heavyweight single-wall boards and lightweight double-wall boards. All boxes had the same dimensions L × W × D of 24 × 12 × 18 inches. ECT was measured using the TAPPI T 839 clamp method, *D* in the MD and CD using the four-point method [27], and caliper with a digital micrometre accurate to 1 μm.

Figure 5.27 shows that the simple McKee formula overestimates the actual McKee, using the full formula McKee brings predicted values closer to the actual. The mean error using the full formula is 0.36 kN, whereas using the simplified formula the mean error is 1.12 kN.

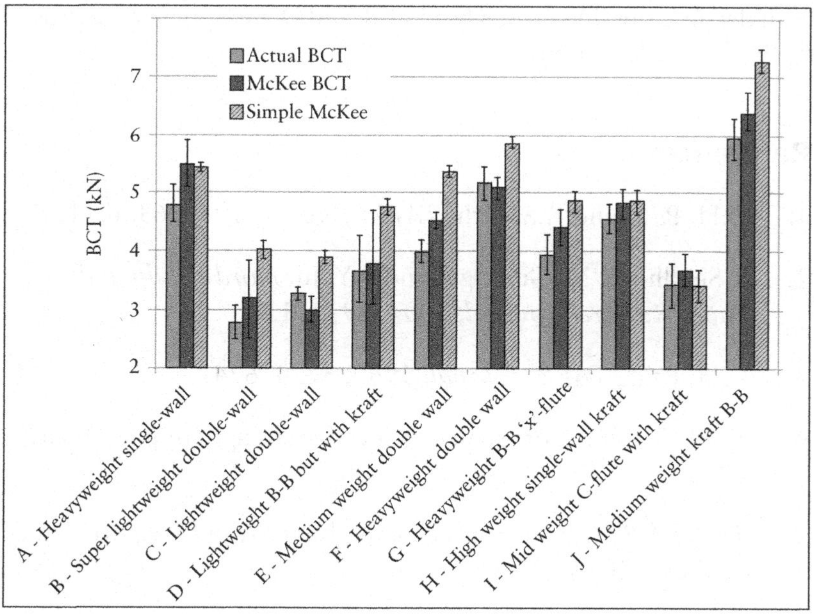

Figure 5.27 Comparison of actual BCT and predicted BCT values using the full and simple McKee formulas for boxes made from the corrugated boards listed in **Table 4.4** (B-B: multiwalled board consisting of 2 B-flute medium layers)

5.8 Summary

The compression strength of paper, board or boxes will combine buckling with compression failure. The SCT test method eliminates buckling from its short-test span, however, this introduces a comparatively high variability in the results which is attributable to paper formation. RCT can be related to SCT using an empirical compression buckling model. ECT will involve the buckling of the entire sample if the height of the test specimen is too high, and will display an out-of-plane buckling pattern on the facings in larger flute (A or C) boards and lighter-weight liners. More accurate ECT values can be predicted and attained if a compressive buckling model is used. Box compression consists of combined compression failure and buckling, and is a better prediction for BCT that agrees with actual values obtained when both ECT and D measurements are used in the full McKee model.

References

1. D.H. Page and R.S. Seth, *TAPPI Journal*, 1980, **63**, 6,113.

2. P. Shallhorn, N. Gurnagul and S.Y. Ju, *Nordic Pulp and Paper Research Journal*, 2004, **19**, 2, 130.

3. D.H. Page, *TAPPI Journal*, 1969, **52**, 4, 674.

4. TAPPI T 826: Short-span compressive strength of paperboard, 2013.

5. TAPPI T 822: Ring crush of paperboard (rigid support method), 2011.

6. R.E. Popil in *Proceedings of the TAPPI PaperCon 2010*, 2–3th May, Atlanta, GA, USA, 2010.

7. J.M. Gere and S.P. Timoshenko in *Mechanics of Materials*, Wadsworth, Belmont, CA, USA, 1984.

8. G. Lindblad and T. Furst in *The Ultrasonic Measurement Technology on Paper and Board*, Lorentzen & Wettre, Kista, Sweden, 2001.

9. R.E. Popil, *BioResources Journal*, 2012, **7**, 2, 2553.

10. P. Shallhorn, S. Ju and N. Gurnagul, *Journal of Pulp and Paper Science*, 2005, **31** 3, 143.

11. R.S. Seth, *TAPPI Journal*, 1985, **68**, 3, 98.

12. TAPPI T 811: Edgewise compressive strength of corrugated fiberboard (short column test), 2011.

13. TAPPI T 839: Edgewise compressive strength of corrugated fiberboard using the clamp method (short column test), 2012.

14. F.J. Plantema in *Sandwich Construction, The Bending and Buckling of Sandwich Beams, Plates and Shells*, John Wiley & Sons, Inc., New York, NY, USA, 1966.

15. R.E. Popil, D.W. Coffin and C.C Habeger, *Appita Magazine & Journal*, 2008, **61**, 4, 307.

16. R.C. McKee, J.W. Gander and J.R. Wachuta, *Paperboard Packaging*, 1961, **46**, 11, 70.

17. W.O. Kroeschell, *TAPPI Journal*, 1992, **75**, 10, 79.

18. L.E. Eriksson, *Containers International*, 1979, **86**, 8, 34.

19. L.E. Eriksson, *Containers International*, 1979, **86**, 9, 64

20. C.J. Wilson, C.J. and B. Frank, *TAPPI Journal*, 2009, **8**, 6, 24.

21. M. Schaepe and R.E. Popil, *Corrugated International*, 2006, September, 3.

22. T.J. Urbanik, *TAPPI Journal*, 1990, **73**, 10, 263.

23. R. Haj-Ali, J. Choi, B. Wei, R.E. Popil and M. Schaepe, *Composite Structures*, 2008, **87**, 4, 321.

24. R.E. Popil and M. Schaepe, *TAPPI Journal*, 2005, **4**, 8, 25.

25. W.J. Whitsitt, *TAPPI Journal*, 1988 **71**, 12, 163.

26. R.C. McKee, J.W. Gander and J.R. Wachuta, *Paperboard Packaging*, 1963, **48**, 8, 149.

27. TAPPI T 836: Bending stiffness, 4 point method, 2016.

6 Testing Methods for Measurement of the Writability and Printability of Papers

6.1 Background

The current popularity of rollerball and fibre-tip pens, which use liquid and gel inks, results in the penetration of inks through the surface and into the sheet. Ink penetrating through pores in the sheet to the other side is called bleed-through and show-through is writing becoming visible on the unwritten side of the sheet, which interferes with the readability of paper on the opposite side. This issue can be ameliorated if there is high paper opacity (the inverse of transparency) obtained by either high basis weight 'β' or the use of high-scattering power fillers such as titanium dioxide (TiO_2). Bleed-through is the result of ink penetrating through the thickness of the sheet and is affected by paper porosity, which is minimised through paper densification *via* a combination of increased wet pressing and high fines content and the use of sizing agents. Both these problems can be reduced in severity by producing paper of higher sheet density, lower porosity and surface roughness, and the addition of surface or internal sizing agents. Show-through can also be limited at a given sheet basis weight by having a high-scattering filler content such as TiO_2 embedded in the sheet.

This chapter describes the development of a test where the volume and rate of application of a non-viscous liquid ink is applied in a controlled and reproducible fashion to paper surfaces. The printed samples are measured on their unprinted undersides for colour changes attributable to the combination of bleed-through and show-through. Quantification of the combined show-through and bleed-through effects using a reproducible, controlled ink-delivery system removes the subjectivity and arbitrariness of visual ranking of handwritten samples using various types of pens and inks. A series of physical tests related to paper surface properties and ink absorption are performed on a sample set for the purpose of developing a predictive model for the writability of paper.

An extensive review of the mechanisms and testing of the wetting of paper surfaces by fluids is contained in the paper by Hubbe and co-workers [1]. It is instructive, for a basic understanding, to examine the Lucas–Washburn equation, **Equation 6.1**, which gives the penetration length 'L' of a column of liquid in a pore or radius 'r' as being the square root of (surface tension 'γ' × cosine contact angle 'θ' × exposure time 't' divided by 'η' viscosity). This is the governing principle of ink transfer to paper, namely, ink is drawn into the paper pores *via* capillary action:

$$L(t) = \sqrt{\frac{r \cos\theta}{2}} \sqrt{\frac{\gamma}{\eta}} \sqrt{t} \tag{6.1}$$

Figure 6.1 is an illustration of the capillary effect, where the fluid surface forms a meniscus with a contact angle determined by the balance between the surface tension of the fluid and the surface energy of the wall material. The term '$\sqrt{\gamma/\eta}$' is the characteristic rate for liquid penetration. Obviously, oil-based viscous inks, as found in ballpoint pens, will penetrate much less than viscous inks such as in gel points, tip markers or fountain pens. Time 't' of ink exposure is governed by the writing or printing speed and is offset by any evaporation or drying. This equation forms the basis for assessing the writability of paper as recently investigated in the MSc thesis of Rioux [2]. The current objective is to describe a means of quantifying the writability of various candidate papers by the measurement of show-through and bleed-through under controlled conditions.

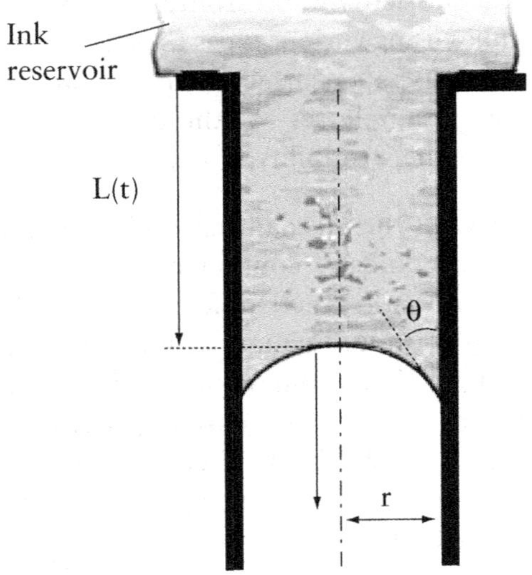

Figure 6.1 Schematic representation of a fluid being drawn into a pore of radius 'r' by capillary action

6.2 Controlled Application of Ink: Using the Bristow Wheel

The standard Bristow wheel apparatus documented by the American Society for Testing and Materials, ASTM D5455 [3], and supported by other literature dating back to 1967 [4], provides a means of applying a swath of ink such that the length of the track is a measure of the absorbency of the sheet on a timescale relevant to handwriting with a pen. The apparatus consists of an accurately controlled revolving wheel onto the periphery of which paper test strips are applied. A slotted die containing a fixed volume of ink is applied to contact the test strip as it approaches to produce a swatch of ink coverage. **Figure 6.2** shows a schematic concept of the Bristow wheel ink application and **Figure 6.3** is a photograph of the apparatus.

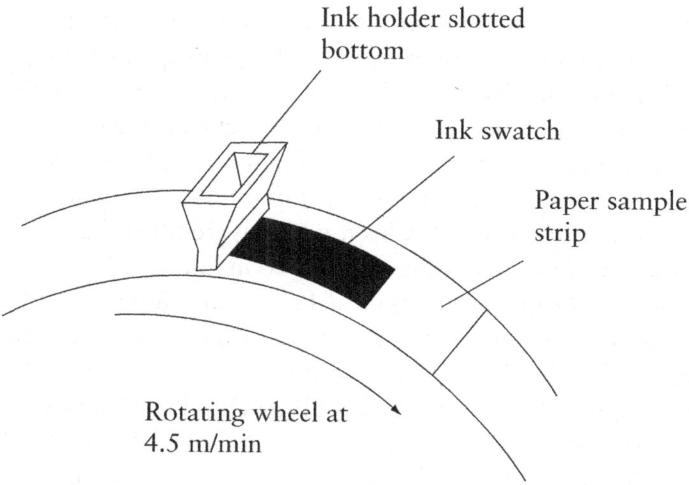

Figure 6.2 The Bristow wheel concept

Figure 6.3 Bristow wheel equipment

A set of 16 commercial notebook papers were supplied by a marketing organisation representing a range of quality and pricing. Printed ruling lines run along the cross direction of paper (CD) of all writing papers and this orientation was checked using the Lorentzen & Wettre (L&W) tensile stiffness orientation (TSO) tester polar plotting capability [5]. Strips of 25 mm width were cut along the CD for use on the Bristow wheel, which were affixed at their ends with adhesive tape. The wheel was set to maintain a linear speed of 4.5 m/min in keeping with typical fast handwriting speed used in other documented studies [6]. The speed is checked by counting the number of revolutions of the rotating drum during a time interval using a stopwatch. Parker Washable Blue Quink™ fountain-pen ink is measured by pipette to 75 µl volume and placed into the slotted die (the slot is 1 × 15 mm). The die is manually lowered into contact with the test strip once it moves into position to print.

The newly inked surface (which is the labelled side of each paper pad) then passes by two 250 W infrared lamps to facilitate faster drying of the surface and removal of the sample. The labelled side of the sample set is the one that is facing the user, once the notebook is opened. Fourdrinier paper machine papers have a felt facing side and a corresponding wire side. The directional drainage of pulp stock during this process produces a fines rich, somewhat denser felt side and a fines depleted, less dense wire side. Liquid absorption can therefore be expected to be side-application dependent. As per usual laboratory practice, six test strips were printed from each sample pad. **Figure 6.4** shows the top side of the printed sample strips. Here, the variation of the length of the track produced is a function of the absorbency of the paper sheet. The variation in optical density or darkness of the strips is dependent on the level of liquid absorption. Intuitively, porous, high bulk, high roughness papers are expected to be highly absorbent and thus unsuitable for writing or printing. However, this expectation may be countered by the presence of sizing agents or surface energy in a sheet, as the penetration of aqueous ink into and through a sheet is governed by capillary action.

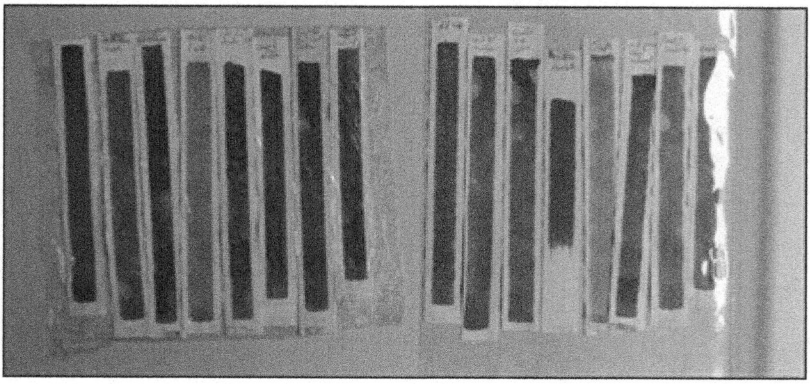

Figure 6.4 Top printed side of the notebook paper sample set using the Bristow wheel

Figure 6.5 shows the undersides of the printed samples displaying varying degrees of show-through and mottle. The variation in bleed-through is dependent on the paper basis weight, density, porosity, and possible presence of sizing and filler. Variation in show-through is attributed to paper opacity, basis weight, sizing and filler. The fifth sample from the right has a very dark and short inked area and a large amount of show-through, indicating that it is highly porous and is not a sized paper. By contrast, the fifth sample from the left has a dark, mottled inked area and comparatively little bleed-through, indicating the presence of a high amount of sizing agent, which limits the penetration of ink from the surface into the bulk of the sheet. What follows is how the show-through/bleed-through of the undersides of the printed samples are quantified.

Figure 6.5 Underside of the inked strips shown in **Figure 6.4** displaying varying degrees of show-through.

6.3 Bleed-Through/Show-Through Measurement

The unlabelled side of the unprinted paper pads were measured for Commission Internationale de l'Eclairage (CIE) Lab* coordinates 'L_u* a_u* b_u*', using a Technidyne Brightimeter S5 with a 0.5" diameter measuring port as per the Technical Association of the Pulp and Paper Industry (TAPPI) method T 524 [7]. Colour and shade can be quantified in terms of these colorimetric values, which incorporates the perception of colour by the human eye. In simple terms, 'a*' describes how 'red' a shade is, 'b*' how 'blue' it is and 'L*' how 'light' it is. Negative values describe the corresponding contrasting qualities, namely '–a*' is green in contrast to red, '–b*' is yellow in contrast to blue and '–L*' how dark in contrast to how light. The scheme to quantify colour and shade is shown pictorially in **Figure 6.6**.

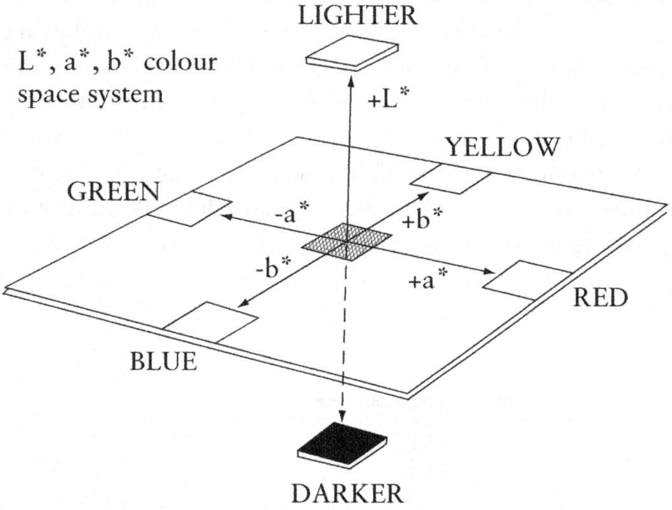

Figure 6.6 Concept schematic for the CIE Lab* colour coordinate system

For many writing and publishing papers, a low 'b*' value is desired which can be obtained by the use of brightening agents or dyes in the pulp. In mechanical pulp grades, such as used in newsprint or softcover books, an increasing 'b*' is a measure of yellowing caused by paper brightness reversion over time. Measurements in the TAPPI T 524 instruments are made using specified directional wide spectrum white light illumination at 45° incidence and the diffusely scattered light emanating at 0° incidence is analysed spectrophotometrically by passing the scattered light through a series of red, blue and green filters onto a photodiode, shown schematically in **Figure 6.7**. The signals obtained by a photodetector behind the alternating red, green and blue filters are processed numerically by the so-called tristimulus functions, which account for the perception colour biases and sensitivities of the human eye. For example, the human eye is least responsive to detecting shades of blue, compared with green or red, and this factor is incorporated into the numerical calculation for L*, a*, b*. The geometry of 45°/0° is considered to emulate a typical reading situation and is sensitive to the machine direction of machine-made paper (MD) or CD orientation of the paper with respect to the plane of light incidence. This method is now widely supplanted by isotropic geometry [8], which makes use of an integrating sphere so that measurements are not directionally dependent, however, the illuminant, detection method and calculations remain the same.

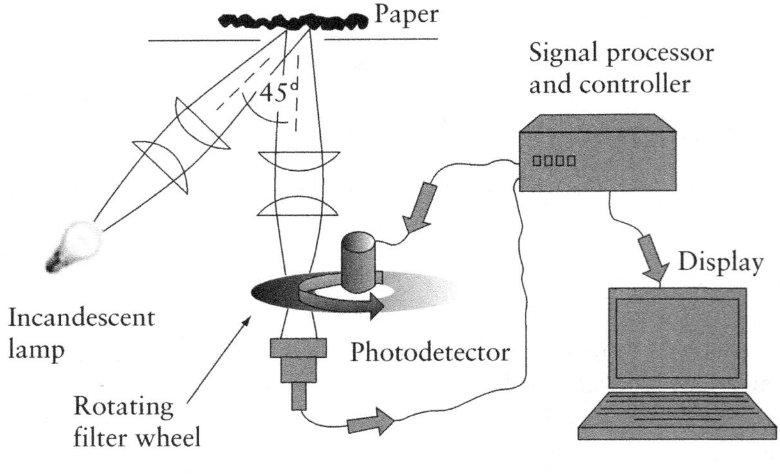

Figure 6.7 Schematic of the geometry used for colour measurement

The method is also widely used to measure the brightness of paper samples, which is the reflectivity of paper at the blue wavelength of 457 nm using broad white light spectrum illumination. Brightness is important in the marketing of publishing writing papers and is sensitive to pulp-bleaching processes, the application of dyes or optical brightening agents.

Figure 6.8 Photograph of the Brightimeter used for colour measurements

The instrument shown in **Figure 6.8** is allowed a 15 min warm-up time after power-up to allow the tungsten halogen lamp to reach full operating temperature, and is then calibrated using a brightness standard tile prior to each use. The undersides of printed samples were measured in three different locations along the length of each strip, and the strips were backed by a pad of unprinted paper. Optical measurements are all made along the CD of the test sample in this TAPPI method. The average of three (Lab)* readings were made for five strips for each sample and an average for the printed underside calculated: L_p^*, a_p^* and b_p^*. The CIE Lab* coordinates were also measured for the corresponding unprinted samples: L_u^*, a_u^* and b_u^*. The change in colour and shade of the underside attributable to printing of the labelled side is called delta 'E' (ΔE), thus:

$$\Delta E = \sqrt{(L_u^* - L_p^*)^2 + (a_u^* - a_p^*)^2 + (b_u^* - b_p^*)^2} \tag{6.2}$$

Equation 6.2 provides quantification of the bleed-through and show-through from inking the samples. Readings were taken using pads of five or more sheets of the unprinted paper, with backing to sheet being measured to take into account the transparency of the paper, which is usually measured as opacity and described in the following.

Opacity *via* method TAPPI T 425 [9] measures the percentage light that is transmitted through a sheet by measuring the amount of light reflected directly back from a sample when backed with a black-absorbing background, compared with when a white background is placed against the sample. A sample with high opacity will reflect much of the light when placed against a black-absorbing background, typically around 90% or more. Highly opaque samples are expected to have small show-through when ink is applied on one surface. The samples that have a high density or have a high amount of filler, such as TiO_2, will have a high opacity and therefore, will be less affected by ink penetrating the surface in terms of show-through to the other side. Considerable commercial research and development is devoted to producing cost-effective paper-filler materials with a high visible light-scattering capability to produce high paper opacity in order to compensate for decreasing basis weight. A Technidyne BNL-3 Opacimeter measured the TAPPI opacity. A single measurement of opacity requires two readings of the sample, one with the instrument's white standard background placed against the back of a sheet facing the light source and optics, followed by another reading with the light-absorbing black-body background placed against the sample.

The instrument calculates opacity as the ratio of the two reflected light levels. Ceramic tiles with calibrated reflectance are used to check the instrument performance after a 15 min warm-up at each use. The brightness, colour, opacity, and so on, of all optical instruments are subject to ageing of the halogen lamps, with spectrometer filter characteristics also changing over time, and so require periodic checking of their performance using brightness and colour standard tiles. Some compensation for ageing can be made by adjustment of electronic gain and offsets prior to the eventual replacement of components.

6.4 Water Drop Contact Angle and Angle Change Rate

For the selected sample set of writing papers, the liquid-absorption properties play a principal role in determining the level of acceptable ink-paper performance. Water drop contact angle and contact angle change rate are introduced to characterise surface-liquid interaction.

The Lucas–Washburn equation (**Equation 6.1**) contains the term *cos* (θ), which is the cosine of the meniscus angle and for a small water drop, of a few microlitres, is also the contact angle the edge of the drop forms with the contacting surface, as shown in **Figure 6.9**. The contact angle is a measure of the wettability of the surface, i.e., a hydrophobic low surface energy will have a high contact angle with a water drop, whereas a highly wettable surface will exhibit a low contact angle. The rate the contact angle changes as the water drop is absorbed into the surface is another useful measure [10].

For the writability evaluation of pen on paper, the water drop contact angle is known to relate to 'ruling quality' such that if the angle the sides of the drop make with the surface are between 110° and 90°, sharp lines are formed because the ink stays where it is applied. Angles greater than 110° will cause applied lines of ink to break up, whereas a contact angle of less than 90° will lead to 'feathering' or lateral spreading of the lines. The change in water drop angle with time is also known to be related to feathering propensity. The delivery of a 4 µl drop, video recording of the water drop absorption and measurement is facilitated by a Fibro PG-3 instrument [10], shown in **Figure 6.10**. Other similar instruments for automated analysis follow the same principles. Water drop contact analysis can also be performed manually albeit tediously, using low-power magnification to observe the droplet contacting the surface, as described in the TAPPI method. The Fibro PG-3 instrument, also known as the FibroDat pocket goniometer, dispenses a microlitre drop onto a paper sample strip, and is equipped with a light source and video camera connected to an analysing computer loaded with the instrument software.

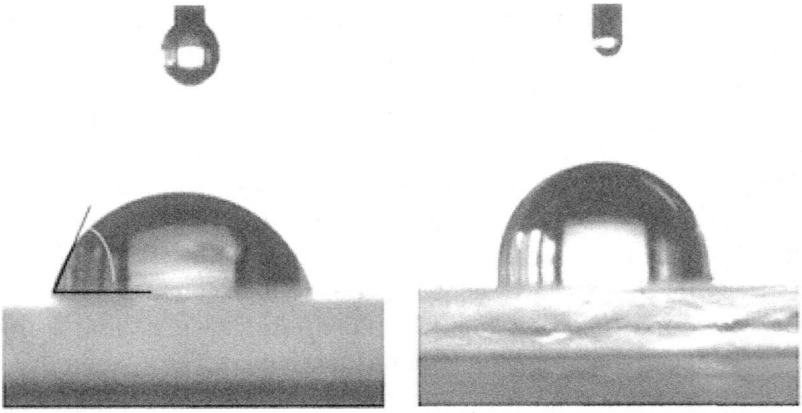

a) b)

Figure 6.9 Low (a) and high (b) contact angle water droplets. The surface in a) is hydrophilic, the angle indicated is <90 ° and in b) it is hydrophobic with a contact angle ≈90°

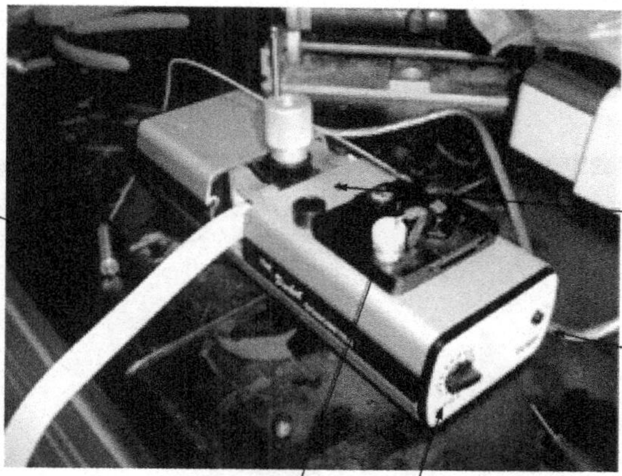

Figure 6.10 FibroDat pocket goniometer for analysing the water drop contact angle (CCD: charge-coupled device, LED: light emitting diode, PC: personal computer and USB: universal serial bus)

Image analysis software is used to analyse the water drop once it is placed onto the surface. The 'dynamic' mode setting allows the recording of the drop becoming absorbed into the sheet surface. A sample screen output is shown in **Figure 6.11a** and shows the screen output at the instant the water drop comes into contact with a copy paper sheet. **Figure 6.11b** shows the drop has been completely absorbed and the contact angle, as a function of time, is displayed. During the analysis of water drops on writing papers, the drops are automatically recorded for 30 s after making contact with the sheet surface. The table of contact angle *versus* time provided by the software is analysed for slope of contact angle change or wetting rate as $\Delta\theta/\Delta t$. A calibration sphere comparable to the water droplet in size is provided with the instrument to ensure, *via* a calibration procedure provided in the software and followed before each use, that the angles and drop volume are correct.

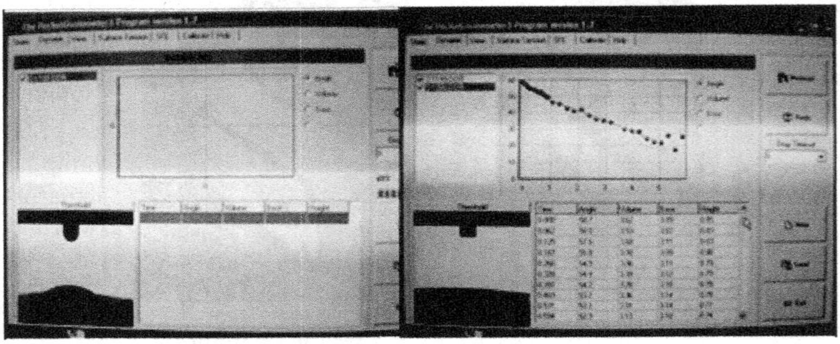

a) b)

Figure 6.11 Sample screen outputs from the FibroDat pocket goniometer for a water drop on standard copy paper. a) Droplet on copy paper data being accumulated and b) final output: contact angle starts at 59° and drops to 20° in 5 s

6.5 Cobb Water Absorption Test

Another test adopted from Swedish Technical Forestry Institute (now Innventia) studies [11] of ink-jet paper characterisation is water absorption (Cobb test) [12], as shown in **Figure 6.12**, where a sample is exposed to a pool of water placed on its surface for 30 s and the absorbed water is measured as a weight difference of the blotted sample after water exposure. The weights of each test piece are recorded to three decimal places on an electronic balance. After 30 s water exposure, samples are removed as quickly as possible from the Cobb tester, placed between kraft pulp blotting papers and subjected to two passes of a 20 kg metal roller prior to weighing. The test piece weight difference before and after water exposure mutiplied by 100 provides the absorbed water in g/m^2. Papers with embedded sizing such as rosin will have low Cobb values, i.e., resisting water absorption, and may be expected to allow less penetration of ink as well.

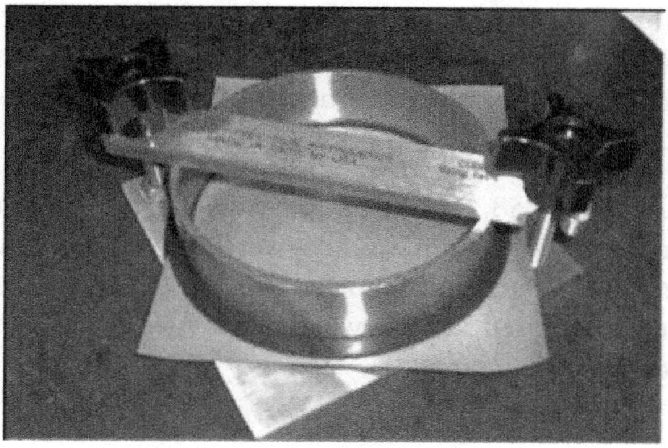

Figure 6.12 A Cobb tester set up ready to test for the amount of absorbed water/area time. Details: 1) rubber pad underneath the sample, 2) rubber O-ring on top of the sample, 3) screw down steel ring tightly, 4) pour 100 cc of 23 °C deionised water and 5) start time

6.6 Caliper and Basis Weight

A dense sheet is expected to be more resistant to ink absorption and penetration therefore, the density of the writing paper sample set is estimated as basis weight divided by the caliper. It is customary to use basis weight in units of g/m^2 and caliper in units of microns with the result of density then being in g/cm^3.

An L&W digital caliper gauge was used to measure 10 repeats per sample to 0.001 mm accuracy. Metal foils and shims of known thickness are used to check the instrument. The method specifies flat parallel steel platens descending onto the specimen at a rate of 4.8×10^{-2} m/min providing, by weight, a pressure of 50 kPa over an area of 2×10^{-4} m^2. The weight and drop rate of the platen takes into account that paper is a viscoelastic compressible material. The diameter of the platen covers the typical topographic waviness of most papers that arises from paper formation or mass non-uniformity. This combination of platen dimensions and clamping pressure as specified above provide somewhat of an overestimate of the caliper, as the surface roughness and topography are not compensated for in this technique.

Punched samples of 53 × 63 mm, as provided by the pneumatic L&W Elmendorf Tear Tester sample punch, were measured to four decimal places for basis weight and reported as g/m^2. Balances are checked by weighing calibrated balance weights. Caliper gauges are checked using metal shim standards.

6.7 Air Permeability or Porosity

Low-density samples which are capable of absorbing a large proportion of liquid are expected to have a high permeability to the flow of air through the sheet. Samples that consist of many pores per unit area, attributable to coarse fibres, low density or a combination of both, will have a high air permeability and so will also probably let a lot of ink through, which contributes to bleed-through and show-through. Air permeability [13] is often synonymous with porosity and is measured routinely in applications where liquid absorption is important, such as the adhesion of fluted medium to linerboard during corrugated board manufacture. Other applications require high air permeability to allow paper webs to adhere well to rolls, e.g., during high-speed printing. An L&W Densometer instrument was used to measure air flow through the samples, which records the number of seconds required for 100 ml of air, at a low pressure of 1.2 kPA, to pass through the sheet. The instrument shown in **Figure 6.13** consists of a free-falling weighted cylinder containing air suspended over a sealing clamping arrangement that allows air to flow through the paper sheet. An optically triggered timer in the instrument records the time required for the top cylinder to drop to the level corresponding to 100 ml of air.

a) b)

Figure 6.13 The air permeability tester (a) and sample clamp detail (b)

The instrument is checked by measuring the time taken for air to flow through a calibrated metal foil orifice, which is supplied by the manufacturer.

6.8 Surface Roughness by Air Leak and Contacting Stylus Profilometer

Surface roughness is generally accepted to affect ink transfer during contacting processes, as the ink film, of limited thickness, applied on the surface cannot reach deep pores. Roughness, related to ink transfer, is commonly assessed indirectly by measuring the amount of air leaking through contacting annular rings [1]. Air escapes through the pores and surface gaps that do not make a sealing contact with the ring cylinder end surface. This method is considered to measure an integral of the volume of pores connected to the surface, based on the geometry of the rings and air pressure difference assuming laminar Poiseuille flow. Therefore, results can be misleading as different paper topographies can produce similar leakage air flow, for example, a few deep pores in an otherwise smooth surface may provide the same air flow as a rough surface with many smaller pores. Nonetheless, the method provides a quick and convenient measurement that is useful for quality control in paper mills

The Sheffield version of the instrument has been applied to predict letterpress solid print density and offset printability using the geometry and parameters shown schematically in **Figure 6.14**, the instrument is shown in **Figure 6.15**.

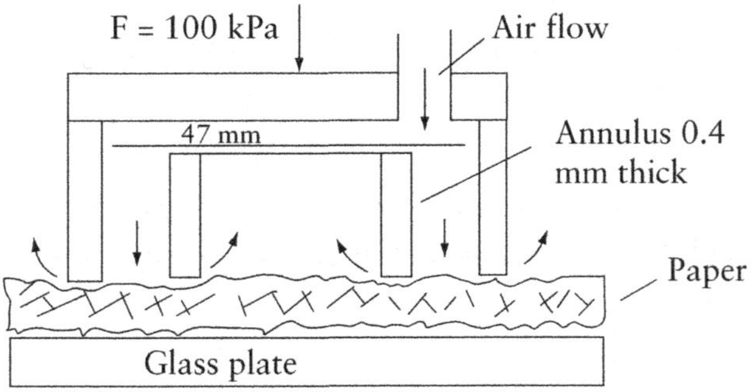

Figure 6.14 Geometry of the Sheffield air leak roughness measurement principle

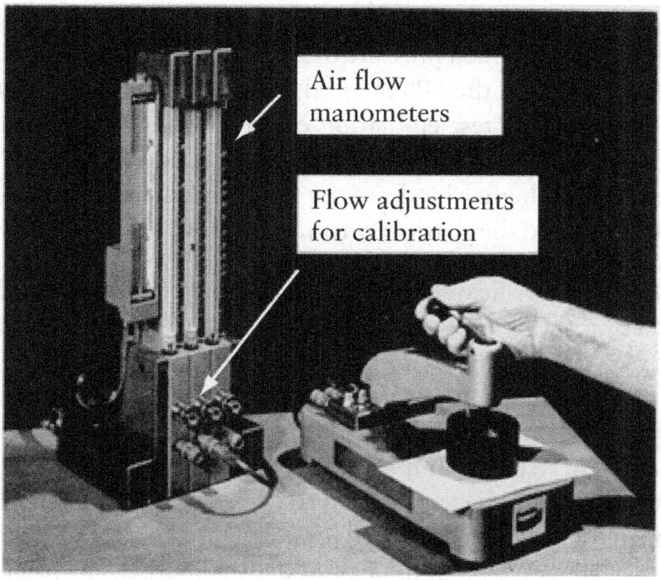

Figure 6.15 Manometer and testing head of the Sheffield roughness instrument

The instrument is calibrated prior to each use by adjusting the manometers using the internal calibration air leaks. Pressure is provided by the weight of the land assembly and the sample rests on a smooth glass plate. Although the manometers read in units of ml/min flow, the conversion to actual air flow is dependent on the range used and is typically 7–9 times the actual manometer read-out, the documented method provides a conversion table if actual air flow is to be recorded.

An alternative more sophisticated form of the air leak method of roughness measurement, considered to correlate with offset dot quality, is the Parker Print-Surf (PPS) [15]. For this method, the thickness of the annulus is decreased from 380 μm in the Sheffield method to 50 μm to correspond to offset printed image dot dimensions. The applied pressure of 2 MPa and a soft rubber backing of the instrument in the PPS method account for the compressibility of the paper under test emulating offset printing conditions. Very smooth surfaces, such as coated printing grades, are also evaluated using a selected higher land pressure on the instrument and a hard backing to extend the sensitivity to variations in the smoother range.

Very rough papers lie outside the range of the air leak methods, and a contacting stylus profilometer provides a useful measurement in these cases. In the author's experience, profilometers intended for use in machine shops, for evaluating ground machined metal surface finishes, correlate well with the PPS S-10 values when evaluating a commercial newsprint sample set with a range of surface roughness. Historically, the Emveco 210-R [16], shown in **Figure 6.16**, was developed to measure linerboard for corrugated boxes and its results correlate well with flexographic print density mottle. The principle of operation is similar to an audio vinyl record phonograph. A projecting tungsten carbide or diamond stylus of radius 25 μm is contained in a 4 N weight paper contacting skid, which is lowered onto a 20 cm paper strip length cut along the MD or CD. Roughness values are higher in the CD than the MD, especially for uncoated papers. The strip is advanced at a speed of 10 mm/s by a drive motor and the stylus with a load force less than 10 mN either protrudes into paper pores or rises at protruding fibre crossings. Consecutive readings are accumulated during a traversing scan and taken at intervals of 250 μm to produce 500 stylus vertical displacement readings along the surface. A 'microdeviation' value is calculated which is found to be sensitive to abrupt changes and less affected by the larger scale wavy topography of paper, as in **Equation 6.3**:

$$\text{microdeviation} = 1{,}000 \times \frac{1}{n-1}\sum_{i=1}^{499}(x_{i+1} - x_i)^2 \qquad (6.3)$$

The calibration uses shims and calculations are in units of thousandths of an inch. Microdeviation values of 50 or lower are considered to be excellent surfaces for subjective print quality, whereas values for untreated linerboard surfaces can be as high as 400. The air leak and profilometer instruments are checked for performance assurance using a smooth (lightweight coated magazine) and a rough paper sample (xerographic copy paper) with recorded historic values.

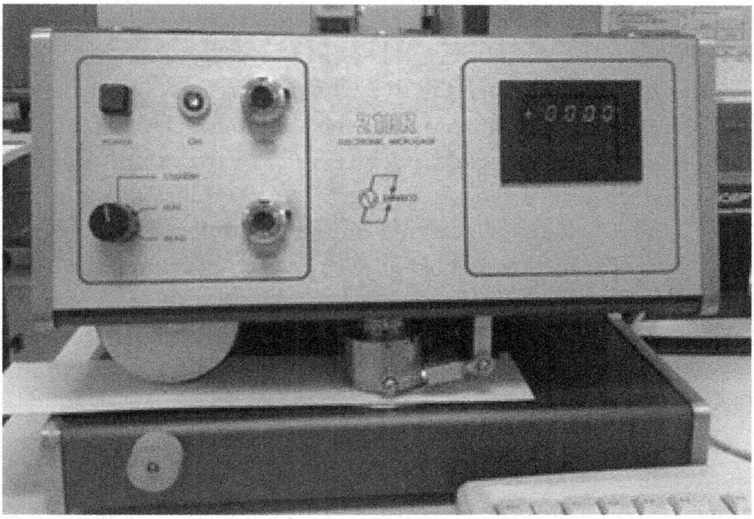

Figure 6.16 Stylus profilometer (Emveco 210-R) developed for the prediction of linerboard printability

The PPS is comparatively more sensitive to high smoothness and requires special attention to ensure that the annular ring lands and the soft or hard rubber backings are clean and free of loose fibres prior to use.

6.9 Sizing Test Ink Penetration – Hercules Size Test

This method, i.e., TAPPI T 530, measures the time taken for a dark dye solution to penetrate to the bottom of the sample once the solution is applied to the top [17]. Sizing agents are often dispersed into the wet-end stock chest of the paper machine to limit the penetration of fluids through the sheet. The change in reflectance of the bottom surface of the sample, which is placed on top of the instrument, is measured to determine the stop time. 10 ml of dye solution is poured into a sample holder, containing the paper sample, and the instrument timer is started simultaneously. A photodiode measures the reflectance of the underside of the illuminated test sample. The timer stops once the reflectivity decreases to 80% of the initial level, i.e., before the dye was poured. Naphthalene B dye solution containing 1% formic acid was used for the Hercules size test (HST), which was arranged to detect 80% of the initial reflectivity level of the bottom of the sample surface; dye solution was applied on the same labelled surface side which was inked by the Bristow wheel on replicate samples. Different reflectance level settings or higher concentrations of formic acid can be used depending on the type of samples under investigation, however, the setting of 80% base reflectivity and a concentration of 1% formic acid are the standard values for evaluating writing or publishing papers which are sized to produce a penetration time of around 180 s. The testing of high basis weight samples may use a concentration of 10% formic acid to limit the time required to reach a 20% decrease in reflectivity.

Prior to each use, the HST undergoes a warm-up period and a reflectivity calibration check using the white and coloured ceramic tiles provided by the instrument supplier. The calibration tiles, sample holder and dye dispenser are shown on the left side of **Figure 6.17**. The procedure ensures that the loss in reflectivity is truly as set by its front panel reflectivity dial indicator, e.g., 80%. This is important to check as the instrument uses incandescent lamps which age. The dye is mixed in equal parts with 2% formic acid and is not kept longer than six months.

a) b)

Figure 6.17 Ink dispenser, clamping sample holder and instrument calibration tiles (green and white) a), and the HST tester b)

6.10 Results and Analysis

Plots of variables reveal relationships that can provide insight into the governing mechanisms that affect writability. The sample set investigated here is a commercial selection of notebook papers representing a range of paper quality and different strategies for limiting ink penetration to the opposite side of the sheet. High sheet density, high basis weight and high surface smoothness, which limit ink penetration physically, may also be balanced by other penetration limiting mechanisms that arise from the use of high-scattering fillers or internal-sizing agents or surface coatings.

Table 6.1 List of commercial notebook writing papers and their physical properties

Sample ID	Visual rank	Writability		Caliper		Basis weight		Sheffield roughness		Porosity	
		ΔE	c.i.	μm	c.i.	g/m^2	c.i.	ml/min	c.i.	s/100 ml	c.i.
D4	4	60.50	5.17	101.8	0.9	73.5	0.3	178.5	14.3	24.1	2.3
D3	4	18.55	1.17	76.2	0.2	57.5	0.2	124.4	22.9	17.1	1.4
D2	2	16.20	4.69	78.7	0.7	56.7	0.5	77.4	21.4	25.1	1.8
D1	2	15.72	1.33	77.5	0.6	55.2	0.6	144.8	19.7	21.1	1.6
C4	3	14.95	3.30	97.9	0.7	75.6	0.5	92	12.8	9.6	1.0
C3	4	14.71	4.51	72.5	1.1	57.1	0.9	56.7	7.4	35.6	3.4
C2	3	11.89	1.18	80.4	0.7	56.2	0.3	41.8	4.5	25.9	2.5
C1	2	10.47	0.83	97.2	1.3	74.8	2.6	147.3	21.4	11.4	1.3
B4	2	9.93	0.51	87.9	1.0	74.2	1.0	50.3	2.3	38.1	1.7
B3	2	8.98	0.46	75.7	0.8	56.5	0.6	97	15.2	19.8	1.8
B2	3	7.90	0.36	102.3	1.8	80.3	0.5	83.8	16.2	35	3.0
B1	1	7.90	0.36	75.5	0.3	51.6	0.2	102.5	8.1	3.9	0.2
A4	4	7.65	0.43	102.5	1.1	77.6	0.5	142.5	12.1	11.4	0.9
A3	1	6.35	0.60	76.1	0.6	58.9	1.0	100	14.4	10.2	0.5
A2	1	5.29	0.17	78.9	0.6	54.4	2.6	131	12.4	23	2.2
A1	1	5.24	0.61	96.8	0.5	80.5	0.2	47.2	6.6	33.8	1.1

Table 6.1 lists the sample set with a subjective visual ranking of the ink show-through on the underside, the writability quantified as the change of colour, 'ΔE' from optical measurements, caliper, basis weight, and air permeability or porosity. Apparent density of the sheet is simply basis weight/caliper. Using units of g/m^2 for basis weight and microns for caliper, yields density in units of g/cm^3. Good writability, meaning a '1' in visual rank or low ΔE, is expected to be associated with a high density leading to a low Sheffield roughness and low porosity.

Figure 6.18 shows that there is not a direct correlation between surface roughness and sheet bulk, which is the inverse of sheet density. A sample set consisting of the same furnish and pulp treatment is expected to have increasing roughness with increasing sheet bulk, as higher wet pressing along with dry-end calendering in paper manufacturing both result in lower bulk and lower smoothness. The scatter of the points in **Figure 6.18** indicate that the relationship between roughness and bulk is altered within the sample set through using different pulp furnishes, fillers or coatings. Sheffield roughness was historically used as a primary predictor of newsprint letterpress solid print ink density. Rougher papers prevented the viscous ink film on the printing plate from penetrating into the sheet surface, resulting in solid print areas with many visually objectionable uninked pores.

Figure 6.19 shows porosity *versus* bulk and many of the samples display the expected trend, i.e., showing a higher porosity with increasing bulk. Sample B1, which is a notable outlier from the trend, has low show-through but a high bulk and low porosity, which suggests that it may be the result of coarse fibres that are highly fibrillated due to refining. The fibre coarseness, which can arise from using softwoods, leads to high bulk but the fines from the fibrillation would impede ink flow-through and lower porosity. In contrast, sample A1, a comparatively expensive premium writing paper, also has little show-through, like sample B1, but has high porosity at low bulk, which suggests that this sample is probably hardwood that is not highly refined but the paper may also contain sizing to limit fluid flow-through the sheet.

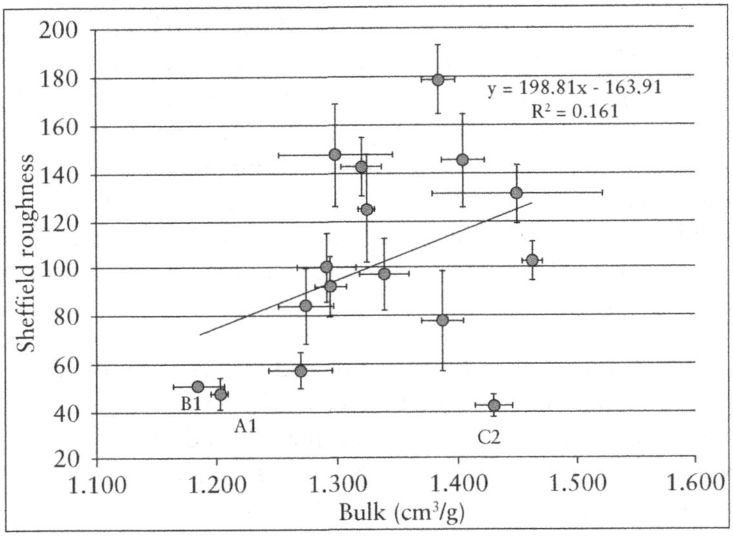

Figure 6.18 Sheffield roughness *versus* bulk (1/density) for the notebook writing paper set (roughness is in Sheffield units, bulk has units of cm^3/g)

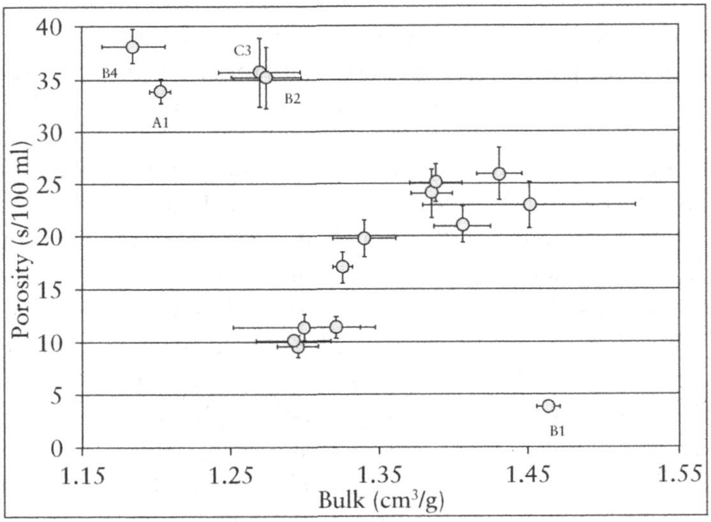

Figure 6.19 Porosity (s/100 ml) *versus* bulk (cm^3/g) of the notebook writing paper sample set

Table 6.2 Measurement results for the commercial notebook paper set

Sample ID	Opacity		Hercules size		Contact angle		Angle rate		$Cobb_{30}$		$Cobb_{30}$/basis wt	
	%	c.i.	s	c.i.	θ°	c.i.	Δθ°/Δt	c.i.	g/m^2	c.i.	$g/m^2/g/m^2$	c.i.
D4	90	0.6	0.5	0.1	56.4	3.0	19.98	0.022	87.3	2.5	1.188	0.084
D3	82.7	0.3	9.5	0.8	91.2	2.0	0.459	0.021	50.1	1.6	0.871	0.028
D2	88	1.0	14.3	2.7	104.5	7.8	0.779	0.035	56.3	1.6	0.993	0.029
D1	87	0.8	15.9	2.0	111.7	2.3	0.282	0.067	33.7	1.8	0.611	0.082
C4	90.4	0.4	5.9	1.3	89.8	3.2	1.302	0.027	79.5	4.1	1.052	0.054
C3	88	0.6	38.2	3.7	106.1	3.5	0.148	0.088	18.8	1.5	0.329	0.027
C2	90.2	0.8	22.5	3.8	108.5	4.2	0.188	0.024	43.9	6.0	0.781	0.107
C1	91.4	0.9	3.3	0.6	85.3	4.2	0.526	0.045	56.7	1.8	0.758	0.035
B4	90.1	0.6	6.7	1.4	78.2	2.0	0.459	0.044	65.4	2.4	0.881	0.035
B3	88.4	0.6	89.3	6.0	110	3.0	0.129	0.032	17.9	1.3	0.317	0.023
B2	92.8	0.2	42.7	5.4	96.3	1.6	0.336	0.051	37.9	2.6	0.472	0.032
B1	84.6	0.3	27.8	2.7	111.8	1.7	0.159	0.084	16.6	0.6	0.322	0.012
A4	90	0.3	23.7	6.6	89.3	4.2	0.231	0.080	45.9	2.9	0.591	0.07
A3	85.6	0.3	118.9	13.9	110.9	3.0	0.173	0.091	16.5	0.3	0.280	0.007
A2	85.3	0.6	110.9	7.4	112.8	3.0	0.103	0.190	14.4	0.7	0.265	0.018
A1	89.4	0.2	225.3	4.1	111.7	1.1	0.198	0.039	19.1	0.5	0.237	0.006

The physical properties of the sample set when examined in relation to writability indicate that the relationships are not straightforward. Although physical properties are expected to influence the absorption of ink flow into a paper, they cannot be used alone to predict performance. Therefore, the absorption of fluid by the paper is examined using various tests, as described earlier and presented in **Table 6.2**.

For the data presented in **Table 6.2,** the liquid-absorption data shows relationships with writability quantified by ΔE. Sample D4 has a ΔE of 60, whereas the rest of the sample set range from 5 to 20 therefore, **Figure 6.20** and **Figure 6.22** plots are semilogarithmic for vertical scale compression.

Figures 6.20–6.22 show the HST, contact angle and the contact angle change rate correlate with ΔE. HST and contact angle change rate are measures of fluid flow into the sheet, whereas contact angle measures how receptive the paper surface is to the aqueous ink.

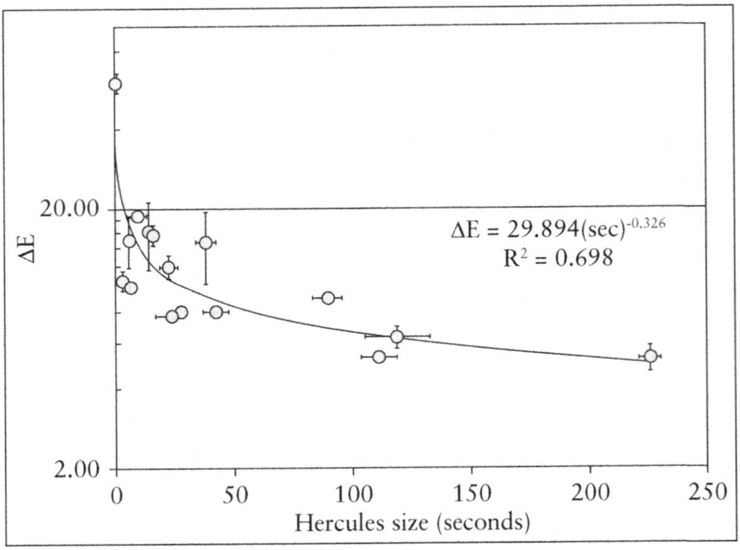

Figure 6.20 ΔE show-through *versus* HST values for the notebook paper sample set

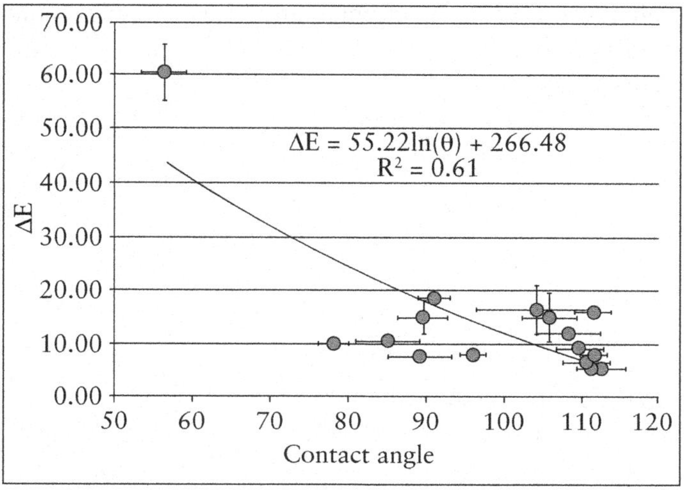

Figure 6.21 Writability or show-through/bleed-through measured as ΔE *versus* the contact angle of ink droplets

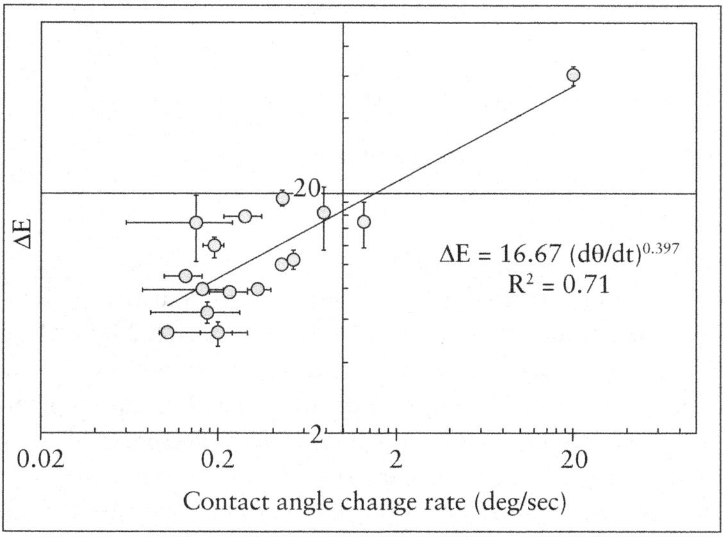

Figure 6.22 ΔE *versus* contact angle change rate for the notebook writing paper sample set

In combination, HST, contact angle and contact angle change rate were used in a best fit model with least error in a linear multiple regression of the form:

$$\Delta E = a_1 x_1 + a_2 x_2 + \cdots + a_n x_n + b \qquad (6.4)$$

The 'regression' function in the Excel Data Analysis Tools package provides a convenient means of doing this. The significance of the fit of the model variables is assessed by an overall high correlation coefficient 'r', low statistical 'p' significance values for the 'x_i' variable coefficients 'a_i' and a low calculated significance 'F' value. In this sample set of notebook and writing tablet papers, the best regression equation was determined to be:

$$\Delta E = -0.0312\,(\text{HST}) + 0.170\,(\theta) + 2.60\,(\text{angle rate}) + 7.86\,(\text{Cobb}_{30}/\beta) - 10.3 \qquad (6.5)$$

with an average error of ± 1.91 points. A comparison plot of the model *versus* actual ΔE values is shown in **Figure 6.23**. A ΔE of 10 points or less is visually acceptable, the model allows a prediction of writability without the need for a printing test. The equation shows that good writability requires a low surface energy (high contact angle), with high wettability corresponding to a high contact angle change rate, good absorption properties (high Cobb/basis weight) and sizing in the sheet (high HST).

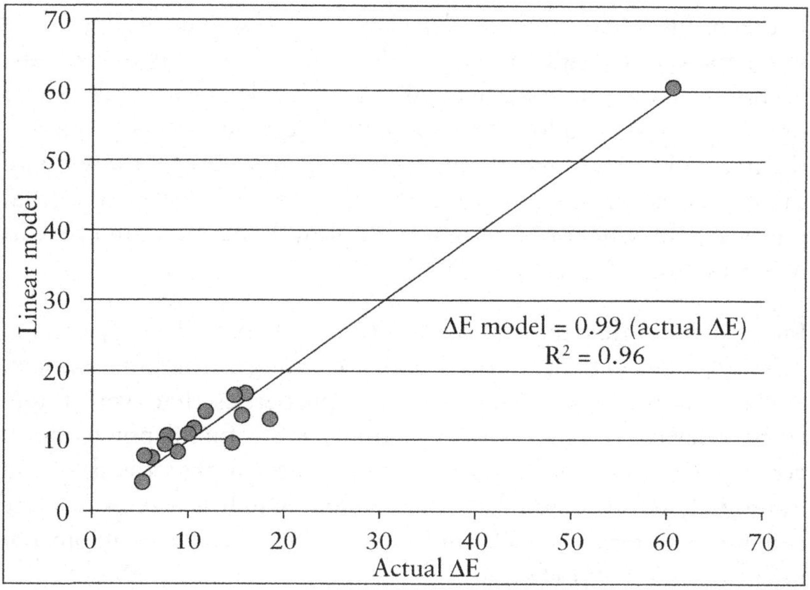

Figure 6.23 Comparison between the linear model and actual ΔE values. Visually acceptable values of show-through have a ΔE value of 10 or less

6.11 Physical Testing for Bank Cheque Ink-Jet Printability

A different set of physical characteristics were found to be of importance for the commercial high-speed ink-jet printing of magnetic ink character recognition font (MICR) on bank cheques. In this case, printed characters along the bottom margin, representing bank account details, must remain precise and intact on the sheet to within specific tolerances determined by MICR reading instruments, such as the Delphax GTX or an RDM Corporation MICR qualifier reader. An acceptable reading from these instruments is a reject value of 5% or less.

If the printing is such that the characters become misaligned or misshapen and result in a poor electronic signal when read, the cheque has to be processed manually incurring delays and costs. Paper properties must be tailored for optimal ink-jet print transfer such that the ink will transfer where intended. Similarly, as was shown with the notebook writing paper set, the objective is to determine what physical tests will correlate best with printability measured here as magnetic ink character percentage (MICR%).

Ink-jet inks in this application are iron oxide particles dispersed in solvents, such as ethylene glycol, and dried using hot air impingement. Surface energy and water-absorption properties that were found to be relevant for notebook writability were found not to be of consequence in this sample set. Data for the sample set is shown in **Table 6.3**, which contains the respective MICR% values, surface roughness in terms of PPS and Emveco 210-R stylus profilometry, microdeviation and HST.

Table 6.3 Results of testing an ink-jet bank cheque printability sample set

Sample ID	MICR%	PPS S-10		Microdeviation		HST	
		μ	c.i.	×10^{-3} inch	c.i	s	c.i.
Best	0.67	5.42	0.22	59.69	5.2	69.4	20.4
Good	0.86	5.23	0.12	54.26	6.32	79.6	27.6
Bad	4.67	4.95	0.19	38.27	6.29	91.7	9
Worst	21	4.5	0.15	33.31	1.6	148.3	20.8

The plot of HST, PPS roughness and microdeviation in **Figure 6.24** show relationships as expected. The roughness, measured by PPS and microdeviation, falls with increasing MICR% misregister, indicating that a rougher surface is better for printed character integrity. This observation of a rougher surface leading to better ink-jet printability is consistent with the conclusions of Lyne and Aspler [18]. Higher HST values correspond to higher MICR% levels, indicating that increased sizing agent is interfering with ink-jet MICR printing. A multiple regression model combining the values from the tests shows the best available predictor to be:

$$\text{MICR\%} = 53.95\,(\text{PPS}) - 0.691\,(\text{microdeviation}) + 0.656\,(\text{HST}) - 296.0 \qquad (6.6)$$

with an average error of +/– 0.07.

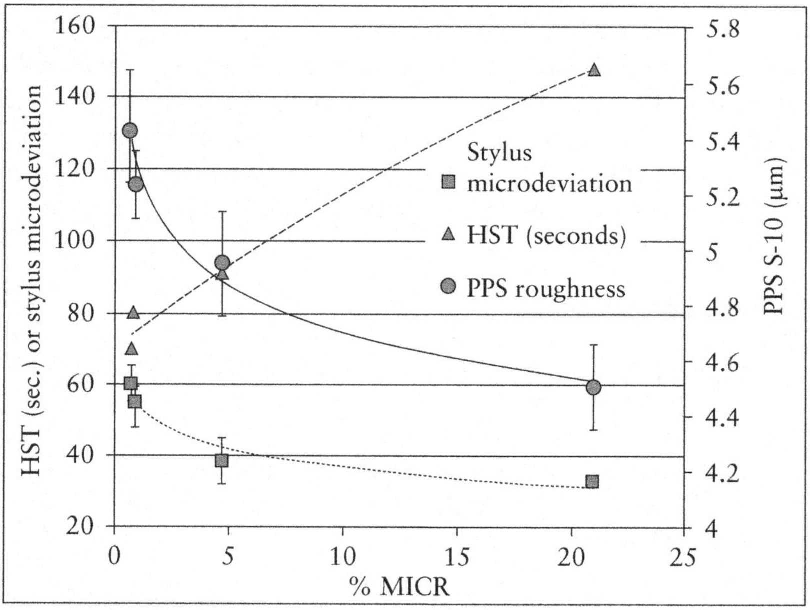

Figure 6.24 HST and surface roughness related to ink-jet printability measured by MICR% misregister

6.12 Summary

How paper reacts to applied liquids determines how writing or printing will appear. Writability or printability can be quantified by optical measurement of inked areas or the undersides of inked sheet surfaces. As paper is comprised of layers of fibres, its porous nature will affect fluid flow. The porosity of paper can be indirectly quantified by measuring low pressure differential air flow through the sheet. Internal-sizing agents limit fluid flow and in the case of low viscosity aqueous inks, improve writability by confining applied ink largely to the paper surface. HST along with the water contact angle relate well to the degree of ink penetration in these cases. Surface roughness is known to affect the transfer of ink films and is indirectly measured by air flow along the surface of a sheet under compression by a stylus profilometer. Linear regression models can be used to combine the results of relevant physical tests to provide a best fit predictive model relating to the printability. Ink show-through for writing papers was found to be related to the combination of HST values, and level of water absorption measured by contact angle change rate and Cobb. Ink-jet printability of solvent-based MICR ink related best to surface roughness and was adversely affected by the presence of sizing. Thus, the measurement of relevant physical properties of paper sample sets can be used to predict their performance for ink to paper applications.

References

1. M.A. Hubbe, D.J. Gardner and W. Shen, *BioResources*, 2015, **10**, 4, 8657.

2. R.W. Rioux in *The Rate of Fluid Absorption in Porous Media*, University of Maine, Orono, ME, USA, 2003. [MSc Thesis]

3. ASTM Standard D 5455-1993: Short-term liquid absorption into paper (Bristow Test), ASTM International, West Conshohocken, PA, USA, 1993.

4. J.A. Bristow, *Svensk Papperstidning-Nordisk Cellulosa*, 1967, 70, 19, 623.

5. G. Lindblad and T. Furst in *The Ultrasonic Measuring Technology on Paper and Board*, Lorentzen & Wettre, Kista, Sweden, 2001.

6. J.M. Hollerbach, *Biological Cybernetics*, 1981, **39**, 2, 139.

7. TAPPI T 524: Color of paper and paperboard 45°/0° geometry, 2013.

8. TAPPI T 527: Color of paper and paperboard d/0° geometry, 2013.

9. TAPPI T 425: Opacity measurement of paper.

10. TAPPI T 558: Surface wettability and absorbency of sheeted materials using an automated contact angle tester, 1997.

11. a) E. Blohm and P. Åslund in *Papers Designed for High Speed Ink-Jet Printing*, Swedish Technical Forestry Institute (now Innventia), Report CW 248, 2004. *http://www.innventia.com/Documents/Rapporter/STFI-Packforsk%20report%20cw248.pdf*

    b) E. Blohm in *High Speed Ink-Jet Printing of Newsprint*, Swedish Technical Forestry Institute (now Innventia), Report CW 119, 2005. *http://www.innventia.com/Documents/Rapporter/STFI-Packforsk%20report%20119.pdf*

12. TAPPI T 441: Water absorptiveness of sized (nonbibulous) paper, paperboard, and corrugated fiberboard (Cobb test), 2013.

13. TAPPI T 460: Air permeability paper, 2006.

14. TAPPI T 538: Smoothness paper and paperboard (Sheffield method) (Revision), 2001.

15. TAPPI T 555: Smoothness paper and paperboard (Print-Surf method) (Revision), 1997.

16. R. Xu, A. Pekarovicova, P.D. Fleming and V. Bliznyuk in *Proceedings of the 2005 TAPPI Coating Conference and Exhibit*, 17–20th April, Toronto, Ontario, Canada, 2005, p.17.

17. TAPPI T 530: Size test for paper by ink resistance (Hercules-type method) (Revision), 2002.

18. M.B. Lyne and J.S. Aspler, *TAPPI Journal*, 1985, 68, 5, 106.

7 'Beater Curves' to Evaluate Pulp Potential: Burst, Tensile, Tear, Opacity

7.1 Background

The potential of pulps for papermaking can be assessed by the preparation of a series of handsheets following the Technical Association of the Pulp and Paper Industry (TAPPI) T 205 method [1]. Many paper mills have their pulps provided as dry sheets which are converted to a dilute stock and processed by refining, a technique borrowed from the food industry to produce starch from grain. Refining wood fibres is achieved by subjecting fibres in aqueous suspension to high shear *via* the mechanical action of the rotating blades. Such action causes fibres to be cut, or more preferably, to become frayed, producing attached and detached filamentary fibrils from the fibre walls. The thrashing of the refining action also causes fibres to be softened producing more collapsible, better bonded fibres, along with a higher density of the sheet due to the fibrils (fines) generated during the process. The effects of refining a pulp for papermaking are assessed by 'beating' the pulps to various levels in a laboratory Norwegian Paper and Fiber Research Institution (PFI) mill device (*Papir- og fiberinstituttet AS*) to achieve varying levels of pulp freeness. The latter term refers to how freely water drains from the resulting pulp. A highly refined pulp will have considerable fibrillation, causing a more tortuous path for water to drain. The Canadian Standard Freeness (CSF), TAPPI T 227 method [2] in fact measures the amount of water that readily drains away from a fixed volume of pulp suspension at a fixed consistency. The physical characteristics of handsheets are measured and related to the level of pulp freeness in order to determine how much refining may be required to reach a certain level of desired end properties. Lower freeness implies the generation of more fines, higher sheet density and shorter fibres, which will result in sheet decreased tear, decreased opacity, but increased burst and increased tensile properties.

Tensile testing was discussed in **Chapter 2**. Opacity, the reflectance of paper at 0° incidence using white light and a back-scatter detector filtered for visible optical wavelength transmission, has been previously mentioned in **Chapter 6** in the characterisation of writing papers. Here, we will cover the popular testing methods of burst and tear of handsheets used to characterise pulps, producing the so-called 'beater curves'.

7.2 Tear Testing of Paper

Curiously, it is the out-of-plane tear test, the Elmendorf test [3] detailed in TAPPI T 414 that is commonly used for paper characterisation, although this mode of tearing is not directly relevant for most end-use fracture properties such as web breaks in a press room. However, it is still regarded as a useful way to rank the relative toughness of different products or variations in quality. The test consists of initiating a crack or a cut along one edge of 4 plies of a rectangular sheet sample. The wire side or top side of the 4-ply sample assembly should all face the same way and 4 plies is the standard assembly. These factors appear to make a difference, as explored by Seth and Blinco [4]. A slitting knife incorporated into the instrument makes an initial cut 20 mm in length along the edge and the firmly clamped 4-ply assembly is then torn out-of-plane by a downward swinging circular sector pendulum attached to one of the test piece clamps. The tear length through the test piece assembly is fixed at 43 mm. The energy expended in the tear is measured by the rise of the pendulum, shown schematically in **Figure 7.1**. The pendulum energy divided by the fixed tear length of 43 mm is reported in mN, generally around 400 to 900 mN for most typical handsheets.

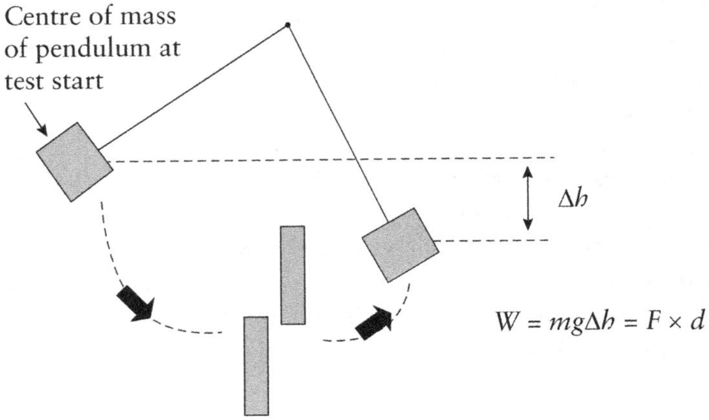

Figure 7.1 Principle of the measurement of the Elmendorf tear test

Figure 7.2 shows a modern digital version of the Elmendorf tester with panel-activated pneumatic sample clamps, pendulum release, digital read-out adjusted for the number of plies being tested, and the ability to calculate and display the average and standard deviation of multiple test results. Pendulum travel in a test is read by an encoder transducer at the pendulum pivot and the particular pendulum type is read by an optical transducer. Different weight pendula are supplied to ensure the instrument is within measuring range for a variety of basis weight papers. In most circumstances, the medium-weight pendulum of 1,600 g will suffice. Specimens are preferably cut with a punch for greatest accuracy, shown in **Figure 7.3**, of dimensions 53 × 63 mm, the length being in the tear direction of interest. Most paper samples produce a slightly higher tear in the cross direction of paper (CD), compared with the machine direction of machine-made paper (MD), meaning it takes more work to cut across the fibres than along the fibres. At least five replicate measurements are made in either the MD or CD to characterise the tear strength for any particular sample. Handsheets have no MD/CD orientation, so five replicate tests suffice.

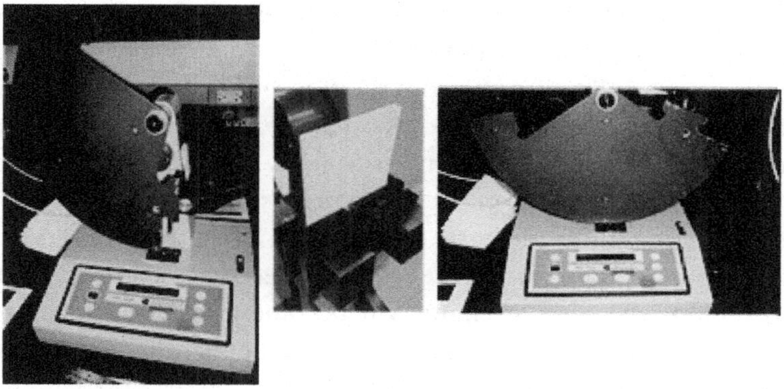

a) b) c)

Figure 7.2 a) A digital Elmendorf tear test unit with its pendulum prior to a test, b) 4 plies of a test sample in the divided clamps and c) pendulum in rest position after a test

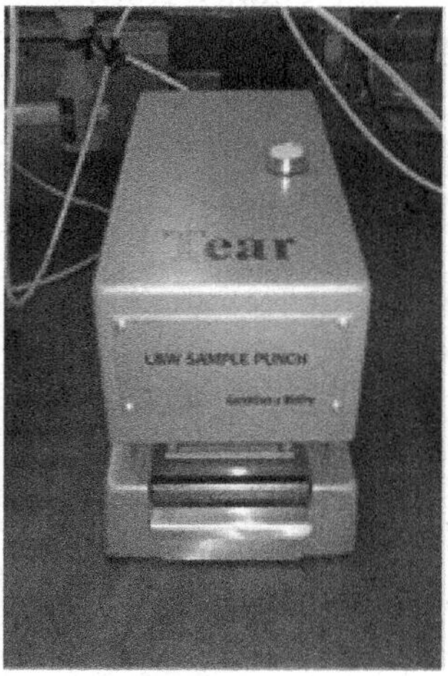

Pneumatically operated punch for tear samples 63 × 53 mm, also useful for quick basis weight measurements using the 5 place balance

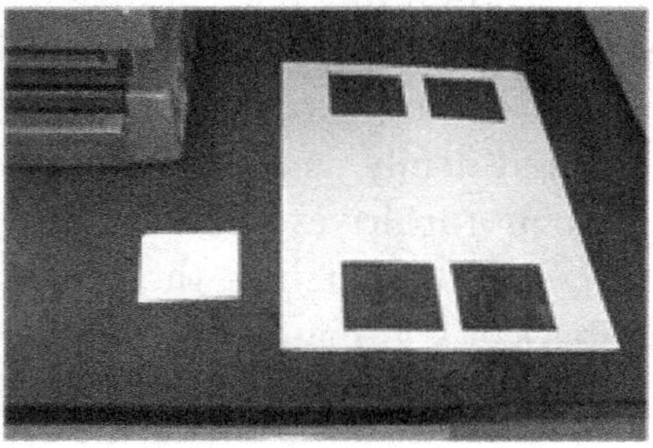

CD tear specimens cut by the punch, the convention is for the length of a sample sheet to be in the MD

Figure 7.3 Punch and orientation of samples for the tear test

The tear test essentially measures the combined work in fracturing and pulling out fibres in the prescribed geometry. Therefore, test values are dependent on the combination of fibre quality and fibre bonding. The fibre-bond strength is usually less than the fibre-fracture strength, which is why in many cases individual fibres can be seen projecting along the edge of a tear. If a process such as refining causes the bond strength to increase, the tear will be initially observed to increase but upon increasing the refining level (shear energy imparted to the pulp) the fibre strength will become affected and the tear strength decreases. What happens for a given pulp at various refining levels in terms of tear strength values, depends on the fibre length, as longer fibres will have more bonds with other fibres, and of course, the fibre strength is largely dependent on the pulp type and wood species used.

7.3 Burst Testing of Paper

One of the most simple but misleading tests for paper is the burst test [5]. Evolved originally from the textile industry and invented in 1887, the test has been used prevalently for quality control in the production of papers made for bags, corrugated packaging and envelopes. Burst strength is the pressure required to puncture a sheet and is measured using a rubber diaphragm secured under an annular ring expanding hydraulically against a paper sample supported firmly against the diaphragm by a circular clamp. An electric motor drives a piston inside a cylinder containing glycerol at the end of which is the rubber sheeting diaphragm, shown schematically in **Figure 7.4**.

A pressure transducer inside the cylinder connected to a peak-and-hold signal digital read-out records peak pressure attained when the sample ruptures as one or more tears. The circular geometry of the test is amenable to mathematical analysis and the main result, pointed out by van den Akker [6], is that the peak rupture pressure 'P' is proportional to the MD and CD tensile strengths $S_{t,MD}$, $S_{t,CD}$, and the square root of the strain to break 'ε_s':

$$P \sim \frac{S_{t,MD} + S_{t,CD}}{2r}\sqrt{\varepsilon_s} \tag{7.1}$$

with '*r*' being the curvature of the diaphragm, which has been experimentally shown to be true. Therefore, the burst test is a combined tensile test in terms that the measured peak pressure at burst is the tensile strength(s) (S_t) multiplied by the square root of stretch to break. The in-plane strain of the paper during the test is proportional to the radius of curvature. Therefore, the side of the sheet facing away from the diaphragm is subjected to a slightly higher strain than the side next to the rubber diaphragm. As the surface structure of many papers is different, especially those made on a fourdrinier sheet paper machine, the test is two-sided. The practice is to test a sheet five times with the sheet wire side facing the diaphragm, then five times again with the top side facing the diaphragm. A sample sheet of any convenient size is placed over the rubber diaphragm and the upper clamp 'tripod' is pneumatically actuated to firmly secure the sample over the diaphragm. A lever is used to manually actuate the piston-geared motor drive and to end the test immediately upon test piece rupture, the instrument is shown in **Figure 7.5**.

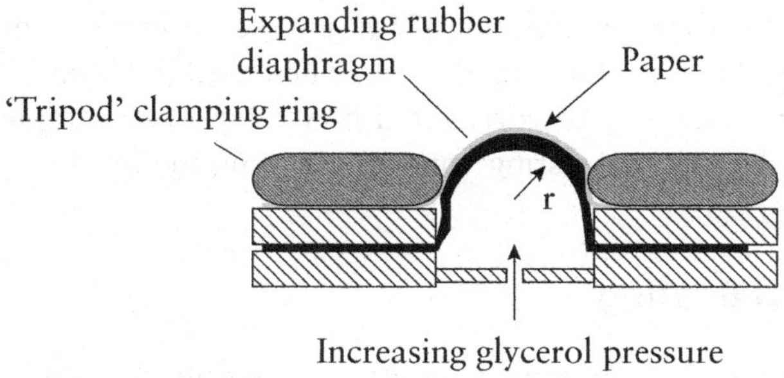

Figure 7.4 Schematic cross-section diagram of the burst test method

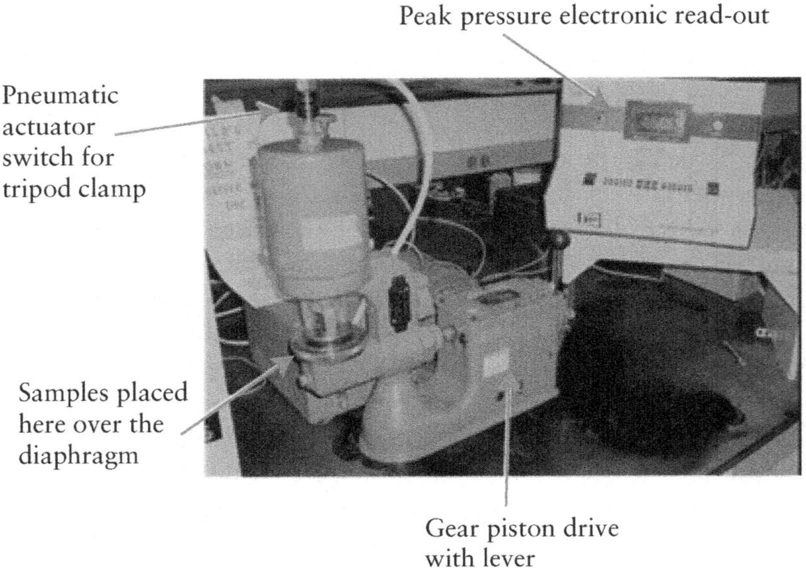

Figure 7.5 A Mullen-style burst tester used for paper and paperboard

Corrugated boards are similarly tested for burst, however, the specification for the test piece clamping is different and uses a larger surface diameter circular ring [7] to ensure no in-plane movement of the sample board during the test. The fluted medium in this case does not contribute to corrugated board burst, so samples can be crushed flat prior to testing without affecting the results.

7.4 Pulp Beating

In this example, a pulp mill submitted three fully bleached softwood kraft samples for evaluation. Bleached hardwood kraft is the most responsive pulp to refining followed by bleached softwood kraft, which is generally about half as responsive, whereas groundwood pulp is least affected. The objective is to attain optimal properties with as little beating as possible to minimise energy costs while optimising quality. The mill produces market pulp in the form of sheets for conversion to other products, such as tissue or writing papers, at other manufacturing facilities. Pulp samples are tested at 10% consistency with up to 30 g placed into a laboratory PFI mill [8], shown in **Figure 7.6**.

Figure 7.6 Photograph of a PFI mill rotor separated from the beater housing

The upper rotor, consisting of chiseled blades, is inserted into the lower housing and rotates at approximately 1,400 rpm while the housing rotates at a lower speed causing a shearing action on the pulp. The number of revolutions are counted and portions of the pulp are removed and diluted to a lower 3% consistency and tested for CSF, which measures the amount of millilitres of water to freely drain from 1 litre of the pulp. The results from beating in the PFI mill and testing the CSF are shown in **Table 7.1**.

Table 7.1 CSF (ml) as a result of PFI mill revolutions for 3 pulp samples			
PFI revolutions	**Sample A**	**Sample B**	**Sample C**
0	748	751	742
1,500	652	632	663
3,000	570	534	544
4,500	413	353	363
6,000	278	230	225

With more PFI revolutions, more work is done on the pulp, which progressively decreases the freeness. Pure water has a freeness of over 900 ml compared with the initial test volume of 1 litre. As the pulp fibres become increasingly swollen and frayed with increased revolutions they are able to retain more free water in the CSF test. The increase in freeness is also expected to translate into increased sheet density due to the increased presence of submillimetre-sized fines and fibrils, and larger strength values in handsheets due to increased bonding as a result of the fibres becoming swollen, softer and less rigid.

The physical properties of the handsheets, due to the pulp being beaten to various freeness levels, are shown in **Table 7.2**. As the number of revolutions increase, the freeness decreases and generally the basis weight of the handsheets increases. The pulp consistency during handsheet making is constant, as per TAPPI T 205, so this observed increase in basis weight with lower freeness is a result of increased retention of the amount of fibre during drainage and more particulate matter in the sheet resulting in higher density.

Table 7.2 Physical properties of handsheets from three pulps beaten to different levels. Number of PFI revolutions are appended to the sample ID letter							
Sample ID		CSF (ml)	Basis weight (g/m²)	Tear index (mN-m²/g)	Burst index (kPa-m²/g)	Opacity (%)	BL (km)
A	0	742	65.5 ± 4.4	15.14 ± 1.02	1.00 ± 0.05	75.7 ± 1.7	2.02 ± 0.08
	1,500	663	68.4 ± 3.9	15.47 ± 0.93	3.63 ± 0.30	66.8 ± 2.2	5.31 ± 0.20
	3,000	554	69.0 ± 3.0	12.50 ± 0.54	4.59 ± 0.25	63.8 ± 1.5	5.93 ± 0.37
	4,500	363	64.1 ± 3.6	11.13 ± 0.62	5.17 ± 0.11	58.1 ± 2.1	7.11 ± 0.22
	6,000	225	65.8 ± 1.9	9.82 ± 0.28	5.36 ± 0.09	60.0 ± 1.4	7.32 ± 0.45
B	0	751	67.0 ± 1.0	12.92 ± 0.19	0.84 ± 0.05	76.4 ± 0.9	1.85 ± 0.05
	1,500	632	69.5 ± 2.3	15.16 ± 0.51	2.83 ± 0.30	59.8 ± 0.6	4.56 ± 0.20
	3,000	534	62.1 ± 2.4	13.12 ± 0.52	3.88 ± 0.32	57.6 ± 2.3	5.11 ± 0.54
	4,500	353	62.0 ± 4.1	11.07 ± 0.74	4.24 ± 0.32	57.6 ± 2.3	6.45 ± 0.58
	6,000	230	63.9 ± 1.6	10.11 ± 0.25	4.64 ± 0.18	59.0 ± 1.5	6.28 ± 0.37
C	0	748	68.0 ± 4.3	10.17 ± 0.64	0.64 ± 0.02	77.8 ± 0.9	1.95 ± 0.17
	1,500	652	69.9 ± 1.2	20.33 ± 0.35	2.47 ± 0.11	58.0 ± 0.6	5.20 ± 0.43
	3,000	570	63.0 ±2.0	15.89 ± 0.50	3.57 ± 0.22	61.4 ± 0.9	6.51 ± 0.68
	4,500	413	60.4 ± 2.1	12.92 ± 0.44	4.40 ± 0.10	56.9 ± 1.9	7.53 ± 0.37
	6,000	278	58.9 ± 1.8	12.31 ± 0.38	4.58 ± 0.24	56.9 ± 1.4	8.19 ± 0.61
BL: Breaking length							

7.5 Results of Physical Properties from Pulp Beating

Of the three pulps, sample C produced the most effective and desirable results. The values of tear, burst and S_t are all proportional to basis weight, so all the strength values are divided by the average basis weight of the handsheets in each sample set. S_t is converted to BL (km) by dividing by the basis weight and multiplying the result by a constant, as shown in **Chapter 2**.

Figure 7.7 shows that the tear index (tear strength divided by basis weight) is improved by 100% with beating to 1,500 revolutions, producing a freeness of 663 ml. Increased revolution decreases the freeness but then the tear index decreases as bonds become stronger but the fibre becomes weaker. Increasing the bonds with increased PFI revolution increases the BL and burst, which have similar trends with CSF, as burst is dependent on S_t. Note that the increase in burst and BL saturate with refining level at a point where the increase in bonds from increased particulate matter becomes offset by shortened and weakened fibres. Smoothed lines drawn through the points in **Figure 7.7** indicate the trends. The beater curve shows that refining beyond a freeness of 660 ml is not worthwhile.

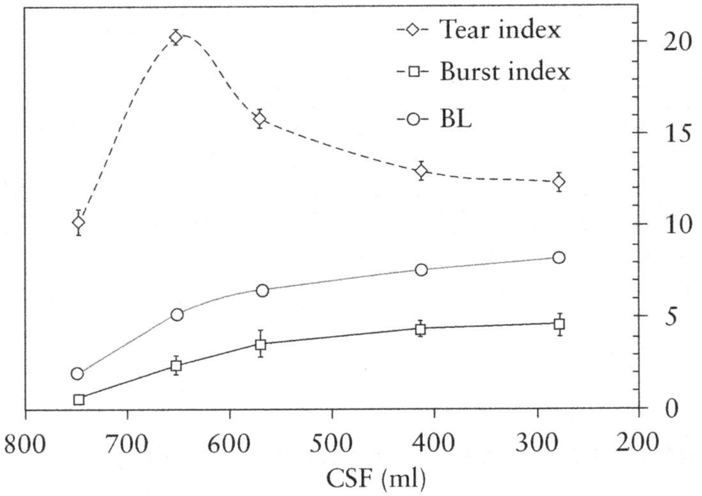

Figure 7.7 'Beater curve' results for pulp sample C [tear index (mN-m^2/g), burst index (kPa-m^2/g) and BL (km)]

The opacity of paper is affected by the combination of the absorption and scattering properties of the fibres, as well as the amount of fibres per unit area or basis weight, which is quantified by the Kubelka–Munk theory [9] for light interacting with paper. Scattering is increased by the number of refractive surfaces per unit volume, so the addition of fines and fibrils to a paper can be expected at first to increase the opacity, and indeed such is the case when a surface coating is applied containing fine small-sized pigment particles of the order of microns or less in diameter. In the case of beaten fibres in wet suspension, the fines and microfibrils bond to paper surfaces and to each other resulting in a higher density but not an increased refractive index gradient, which would otherwise increase the scattering power. Much like the compaction of snow on roads underneath tyre tracks reduces the opacity such that the underlying pavement shows through the overlying snow, the increase of fibrils and fines and increased fibre bonding all result in a higher sheet density that occurs with lower freeness causing a loss in paper opacity. The linear dependence of opacity on basis weight is removed by dividing opacity values to calculate the opacity index, shown as a function of freeness for the three pulp samples in **Figure 7.8**.

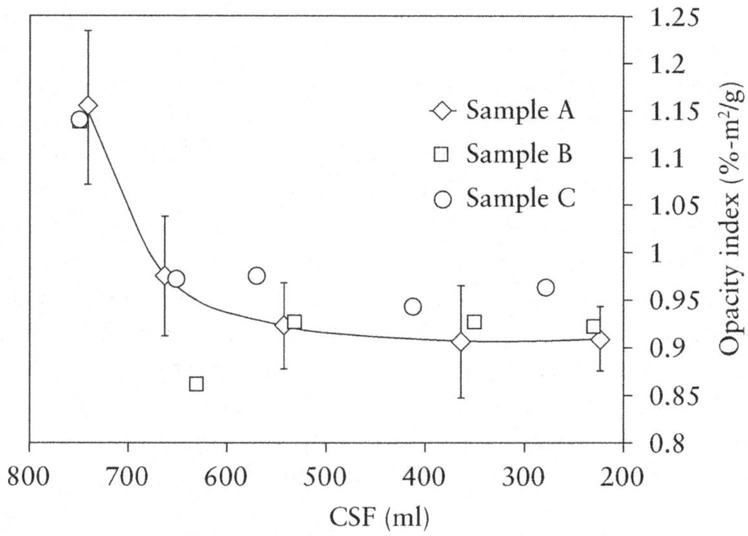

Figure 7.8 Opacity index for all three pulp samples *versus* freeness

All three pulp samples show a similar declining opacity trend with decreasing freeness, consistent with the concept of loss of scattering due to increased sheet density. Similar losses in opacity from paper densification affect the measurement of paper formation (areal mass uniformity) when lightweight paper is calendered to attain caliper and surface smoothness [10]. This effect, which has led to misleading attempts of formation optimisation in mills, prompted the development of alternative paper formation measurement methods. These include imaging of the transmission of electrons from radioactive sources or electron microscopes or long wavelength X-rays [11] to avoid the optical scattering effects that occur otherwise when using visible light transmission for paper imaging.

7.6 Summary

Pulps are often evaluated for their response to refining using a laboratory beater, which is used to prepare samples with a progressive lower amount of freeness. A lower freeness pulp is the result of the amount of refining energy that has been imparted from beating, resulting in higher water retention in the pulp when drained through a screen. Higher sheet density resulting from softer more collapsed fibres and the presence of fines and microfibrils creates more stress transfer between fibres when the sheet is under load so that S_t increases with refining work. The common 'beater curve' follows pulp development with increasing beating work by measuring the out-of-plane tear and tensile or burst properties. Increased fibre bonding with refining level causes the tear strength to increase with some refining but with further refining is observed to decrease, while the S_t steadily increases with refining. The objective of the 'beater curve' is to determine what freeness level is required to obtain the optimal balance between tear strength and S_t.

If other properties were to be followed with changes in pulp freeness, the expectation from first principles emphasised in this book, is that the elastic modulus, measured either mechanically or by ultrasonic propagation, will be observed to follow the same trend as either

S_t or burst strengths. Compression strength measured by the short-span method is also expected to increase with refining work based on the tensile–compression strength relationship and their mutual dependence on elastic modulus. Bending stiffness, being $Et^3/12$, should be expected to decrease as although modulus 'E' increases with sheet density, the corresponding decrease in caliper from softer more collapsible fibres and higher bonding will make the handsheet less structurally rigid.

Increased sheet density with refining produces less scattering, resulting in a loss of opacity and decreased brightness, an undesirable result for writing or printing papers. Since most paper products are required to meet stringent marketing specifications in terms of S_t or tear strength properties, opacity or brightness, it is obvious how the combined engineering and balancing of all these interrelated and interdependent properties in the papermaking production process is often deservedly called an art. It is an art that is certainly amenable to physical testing measurement and quantified characterisation.

References

1. TAPPI T 205: Forming handsheets for physical tests of pulp (Revision), 2012.

2. TAPPI T 227: Freeness of pulp (Canadian standard method) (Revision), 1999.

3. TAPPI T 417: Internal tear resistance of paper (Elmendorf-type method).

4. R.S. Seth and K.M. Blinco, *TAPPI Journal*, 1990, 73, 1, 139.

5. TAPPI T 403: Bursting strength of paper (Revision), 1997.

6. J.A. van den Akker, *Research Bulletin*, 1938, 4, 3, 46.

7. TAPPI T 810: Bursting strength of corrugated board, 1985.

8. TAPPI T 248: Laboratory beating of pulp (PFI mill method), 2015.

9. J.S. Arney, J. Chauvin, J. Nauman and P.G. Anderson, *Journal of Imaging Science and Technology*, 2003, **47**, 4 339.

10. A. Komppa and K. Ebeling in *Proceedings of the* 7th *Role of Fundamental Research in Papermaking Symposium*, September, Cambridge, UK, 1981, Mechanical Engineering Publications, London, UK, 1983, p.603.

11. H. Tomimasu, D. Kim, M. Suk and P. Luner, *TAPPI Journal*, 1991, **74**, 7, 165.

Abbreviations

%M	Moisture content percentage in paper
ASTM	American Society for Testing and Materials
B-B	Multi-walled board consisting of 2 B-flute medium layers
BCT	Box compression test strength
BL	Breaking length
BW	Backing wire
CAD	Computer-aided design
CCD	Charge-coupled device
CD	Machine direction of paper
CIE	Commission Internationale de l'Eclairage
CMT	Corrugated medium test
CSF	Canadian Standard Freeness
cv	Coefficient of variation
CW	Conveying wire side of the paper
D	Bending stiffness(es)
E	Elastic modulus(i)

ECT	Edge compression strength test
FEA	Finite element analysis
GSM	Grams per square metre
HST	Hercules size test
ID	Identification
IPST	Institute of Paper Science and Technology
L&W	Lorentzen & Wettre
MD	Machine direction of machine-made paper
MICR	Magnetic ink character
MICR%	Magnetic ink character percentage
MSE	Mean square-root error
N&W	Noble and Woods
PC	Personal computer
PFI	Norwegian Paper and Fiber Research Institute
PM	Paper machine
PPS	Parker Print-Surf
RBA	Fibre-relative bonded area
RCT	Ring crush test
RH	Relative humidity
RT	Room temperature
S_b	Tensile stiffness(es)

SCT	Short-span compression test or strength
S_t	Tensile strength(s)
STFI	Swedish Technical Forestry Institute, now Innventia
TAPPI	Technical Association of the Pulp and Paper Industry
TIO_2	Titanium dioxide
TSI	Tensile stiffness index
TSO	Tensile stiffness orientation
USB	Universal serial bus
UTM	Universal testing machine
ZD	Out-of-plane z direction of paper

Index

A

B

C

D

E

F

G

H

I

J

K

L

N

O

P

Q

R

S

T

U

V

W

X

Y

Made in the USA
Monee, IL
08 November 2026

67816259R00136